Pesticides
Methods for Their Residues Estimation

NIPA® GENX ELECTRONIC RESOURCES & SOLUTIONS P. LTD.
New Delhi-110 034

Pesticides

Methods for Their Residues Estimation

Dr. Beena Kumari
Senior Analytical Chemist
Department of Entomology
Ch. Charan Singh Haryana Agricultural University
Hisar, Haryana

and

Dr. T.S. Kathpal
(Retd.) Senior Pesticide Chemist and
Former, Professor & Head
Department of Chemistry & Biochemistry
Ch. Charan Singh Haryana Agricultural University
Hisar, Haryana

NIPA® GENX ELECTRONIC RESOURCES & SOLUTIONS P. LTD.
New Delhi-110 034

NIPA® GENX ELECTRONIC RESOURCES & SOLUTIONS P. LTD.

101,103, Vikas Surya Plaza, CU Block
L.S.C. Market, Pitam Pura, New Delhi-110 034
Ph : +91 11 27341616, 27341717, 27341718
E-mail: newindiapublishingagency@gmail.com
web: www.nipabooks.com

For customer assistance, please contact
Phone: + 91-11-27 34 17 17
Fax: + 91-11- 27 34 16 16
E-Mail: feedbacks@nipabooks.com

ISBN: 978-81-19254-93-4

Composed and Designed by NIPA®.

Foreword

Agriculture is the back bone of Indian economy; hence agricultural development is central to all strategies for planned development. For attaining self sufficiency in food and fibers in India, pesticides have been playing a vital role for the last 4-5 decades. Consequently, pesticides have become one of the most important components of modern agricultural technology and are increasingly becoming indispensable for the successful control of crop pests and vectors of human and animal diseases. Hence, the option for intensive agricultural plant protection methods will continue until a time when an alternative to this is invented. Presently, 221 pesticides are registered for agricultural use in India. The use of pesticides has been a mixed blessing. On one side, these have helped in enhancing and stabilising agricultural production, on the other side, their widespread indiscriminate use has given rise to many problems viz., persistence of toxic residues in the environment and food leading to human health hazards, development of resistance in insect-pests and resurgence of pests. For these reasons, environmentally persistent pesticides like DDT, HCH, aldrin, dieldrin, endrin, chlordane, and toxaphene have been banned globally. Newer more potent and easily biodegradable molecules have replaced the persistent chemicals resulting in reduction of pesticide consumption in the recent past. However, increased public awareness and entrance into world trade organisation (WTO) era have generated keen interest in quality produce as well as residue free food commodities. Although new generation pesticides are moderately persistent, after performing their action, these may leave toxic metabolites and other converted moieties, collectively termed as residues, in crop products, soil, water and air. Therefore it is imperative to know the quality and quantity of these moieties in different components of the environment including food commodities for which sensitive, authentic and rapid analytical techniques are the pre requisite.

Pesticide residue analysis is a highly technical area of research involving costly solvents and sophisticated equipments like GC, GCMS and HPLC. In the era of rapidly developing sensitive and sophisticated techniques for the quantitative evaluation up to nanogram or picogram levels of pesticide residues, the students and scientists/ analysts still have to rely on the easily readable and comprehensive subject materials. Detection and quantification of multi-residues in the crop produce is another challenge to an analyst. This book dealing with pesticides and different methods for estimation of single pesticide residue as well as multi- residue analysis is one such effort by the authors who have made industrious efforts and collected exhaustive information on the subject in a systematic and lucid manner. The book includes maximum residue limits (tolerance levels) fixed for different pesticides for different food commodities and glossary of terms generally used in the field of pesticides. I am sure this book will serve as a reference guide for the analytical chemists. I compliment the authors for this contribution in the field of pesticide residue analysis.

(R.P. Narwal)
Director of Research
CCS Haryana Agricultural University
Hisar

Preface

Pesticides are an important component in pest management strategies for food production and public health. Regardless of the method of application, pesticides ultimately reach the soil, which serves as a reservoir for these chemicals. Pesticide in the parent form or in its converted form having toxicological significance is collectively termed as pesticide residues. Pesticide residues are absorbed by plants, enter the food chain and accumulate in human as well as animal body fat. Some residues are carried away by wind in the vapor form to long distances away from the source and then get collected by dust and rain before they are deposited on the ground again. Injudicious and indiscriminate use of these chemicals is responsible for environmental pollution and non-target toxicities. Many problems like persistence of toxic residues in the environment, development of resistance in insect-pests and resurgence of pests are also associated with these toxic chemicals. Pollution of environment poses a threat to the health and wealth of every nation. Consequently it is essential to monitor the levels of organic pollutants in the environment. The problem has further been complicated due to presence of multi-residues in environmental components and food commodities. Residues are inevitable by- product of pesticide use. With the benefits of increased knowledge and experience, it is apparent that these products must no longer be used as they were in the past. All these issues have changed pest control from a simple task in the old days into the complex and publicly-sensitive operation of today. People who develop and supervise modern pest control methods must be highly trained in many areas of pesticide usage. The fact regarding pesticide residues that they are found in all components of the environment is only due to the significant advances in analytical chemistry. The techniques are now so sensitive that detection level that can be reached is equivalent to detecting one teaspoon of salt in one million gallons of water. Residue levels even lower than this can also be detected. The knowledge on the fate of pesticides, after application in the environment is useful to research workers associated with their safe and effective use. It is important to know whether a xenobiotic will persist or it will be inactivated. Frequently persistence is desirable for sustained control. However, for many cases, rapid detoxification of the compound may be desirable to avoid cumulative effects.

In recent years, pesticide residue analysis has become the world fastest growing and most rapidly changing subject. Pesticide residue analysis is a highly technical area of research involving costly solvents and sophisticated equipments like GC, HPLC, HPLC-MS, GC-MS etc. Only highly trained personnel can conduct the sensitive analysis of nano and picogram quantities of pesticide residues present in different matrices. Through this book, an attempt has been made to give comprehensive account of methods that are presently being followed for sampling, extraction and estimation of pesticide residues employing different analytical techniques.

This book strives to highlight the traditional approaches of sample preparation of organic samples that have been, and continue to be, used whilst also considering modern alternatives. It is compiled in 11 chapters. Chapter 1 and 2 explain about the brief history of pesticides development, present day pesticides and metabolic pathway of selected pesticides from different groups of pesticides. Chapter 3 illustrates the sampling, extraction, clean-up and solid phase extraction for residue analysis in detail. Different analytical techniques like TLC, GLC, HPLC, super critical fluid chromatography (SFC) and HPTLC, have also been included through chapters 4, 5, 6, 7 and 8. Maximum residue limits (MRL) have been explained in chapter 9. Recently developed multiresidue methods have been outlined in chapter 10. An extensive list of glossary is provided in chapter 11 in order to make information on pesticides and their residues more widely available. Help has been taken from a number of standard reference books and research papers to enrich the knowledge. We hope this book will provide students with sufficient background for the pesticides, their metabolism and above all various techniques for the estimation of residues of the toxic xenobiotics. I also hope that the book will prove valuable to all those who are engaged in the area of pesticides.

Authors

Contents

Chapter 1
Introduction

Part A

1. Pesticides

A pesticide is defined as any substance intended for preventing, destroying, attracting, repelling or controlling pests including unwanted species of plant, or animals during the production, storage, transportation, distribution and processing of the food, agricultural commodities or animal's feed or which may be administered to the animals for the control of ectoparasites. Pesticides include insecticides, herbicides, fungicides, acaricides, nematicides, etc.

According to food and agricultural organization (FAO) 1986, the term pesticide has been defined as "any substance or mixture of substances intended for preventing or controlling any pest and includes any substance or mixture of substances intended for use as plant growth regulator (PGR), defoliant or desiccant. The term excludes fertilizers and antibiotics or other chemicals administered to animals to stimulate their growth or to modify their reproductive behaviour". Pesticides are the largest group of poisonous substances that are widely broadcast today.

1.1 Type of Pesticides

Pesticides are often referred to according to the type of pest they control. Another way to think about pesticides is to consider those that are chemical pesticides or are derived from a common source or production method. Other categories include biopesticides, antimicrobials, and pest control devices.

1.1.1 Chemical Pesticides

Some examples of chemically-related pesticides follow. Other examples are available in sources such as Recognition and Management of Pesticide Poisonings.

1.1.1.1 Organochlorine Insecticides

These were commonly used in the past, but many have been removed from the market due to their health and environmental adverse effects and their persistence (e.g. DDT and chlordane).

1.1.1.2 Organophosphate Pesticides

These pesticides affect the nervous system by disrupting the enzyme that regulates acetylcholine, a neurotransmitter. Most organophosphates are insecticides. They were developed during the early 19th century, but their effects on insects, which are similar to their effects on humans, were discovered in 1932. Some of them are highly poisonous and were used in World War II as nerve agents. However, they usually are not persistent in the environment.

1.1.1.3 Carbamate Pesticides

They affect the nervous system by disrupting an enzyme that regulates acetylcholine, a neurotransmitter. The enzyme effects are usually reversible. There are several subgroups within the carbamates.

1.1.1.4 Pyrethroid Pesticides

These pesticides were developed as a synthetic version of the naturally occurring pesticide, pyrethrin, which is found in chrysanthemum flowers. They have been modified to increase their stability in the environment. Some synthetic pyrethroids are toxic to the nervous system.

1.1.2 Biopesticides

Biopesticides are certain types of pesticides derived from such natural materials as animals, plants, bacteria, and certain minerals. For example, canola oil and baking soda have pesticidal applications and are considered biopesticides. Biopesticides fall into three major classes:

1.1.2.1 Microbial Pesticides

They consist of a micro-organism (e.g., a bacterium, fungus, virus or protozoan) as the active ingredient. Microbial pesticides can control many different kinds of pests, although each separate active ingredient is relatively specific for its target pest[s]. For example, there are fungi that control certain weeds, and other fungi that kill specific insects. The most widely used microbial pesticides are subspecies and strains of *Bacillus thuringiensis*, or Bt. Each strain of this bacterium produces a different mixture of proteins, and specifically kills one or a few related species of insect larvae. While some Bt's control moth larvae found on plants, other Bt's are specific for larvae of flies and mosquitoes. The target insect species are determined by whether the particular Bt produces a protein that can bind to a larval gut receptor, thereby causing the insect larvae to starve.

1.1.2.2 Plant-Incorporated-Protectants (PIPs)

These are pesticidal substances that plants produce from genetic material that has been added to the plant. For example, scientists can take the gene from the Bt pesticidal protein, and introduce the gene into the plant's own genetic material. Then the plant, instead of the Bt bacterium, manufactures the substance that destroys the pest.

1.1.2.3 Biochemical Pesticides

They are naturally occurring substances that control pests by non-toxic mechanisms. Conventional pesticides, by contrast, are generally synthetic materials that directly kill or inactivate the pest. Biochemical pesticides include substances, such as insect sex pheromones, which interfere with mating, as well as various scented plant extracts that attract insect pests to traps. Because it is sometimes difficult to determine whether a substance meets the criteria for classification as a biochemical pesticide, environmental protection agency (EPA) has established a special committee to make such decisions.

1.2 Pest Types

Pesticides that are related because they address the same type of pests include:

1.2.1 Algicides

Control algae in lakes, canals, swimming pools, water tanks, and other sites.

1.2.2 Antifouling Agents

Kill or repel organisms that attach to underwater surfaces, such as boat bottoms.

1.2.3 Antimicrobials

Kill microorganisms (such as bacteria and viruses).

1.2.4 Attractants

Attract pests (for example, to lure an insect or rodent to a trap). (However, food is not considered a pesticide when used as an attractant).

1.2.5 Biopesticides

Biopesticides are certain types of pesticides derived from such natural materials as animals, plants, bacteria, and certain minerals.

1.2.6 Biocides

Kill microorganisms.

1.2.7 Disinfectants and Sanitizers

Kill or inactivate disease-producing microorganisms on inanimate objects.

1.2.8 Fungicides

Kill fungi (including blights, mildews, molds, and rusts).

1.2.9 Fumigants

Produce gas or vapor intended to destroy pests in buildings or soil.

1.2.10 Herbicides

Kill weeds and other plants that grow where they are not wanted.

1.2.11 Insecticides

Kill insects and other arthropods.

1.2.12 Miticides (also called acaricides)

Kill mites that feed on plants and animals.

1.2.13 Microbial Pesticides

Microorganisms that kill, inhibit, or out compete pests, including insects or other microorganisms.

1.2.14 Molluscicides

Kill snails and slugs.

1.2.15 Nematicides

Kill nematodes (microscopic, worm-like organisms that feed on plant roots).

1.2.16 Ovicides

Kill eggs of insects and mites.

1.2.17 Pheromones

Biochemicals used to disrupt the mating behavior of insects.

1.2.18 Repellents

Repel pests, including insects (such as mosquitoes) and birds.

1.2.19 Rodenticides

Control mice and other rodents.

The term pesticide also includes these substances:

1.2.20 Defoliants

Cause leaves or other foliage to drop from a plant, usually to facilitate harvest.

1.2.21 Desiccants

Promote drying of living tissues, such as unwanted plant tops.

1.2.22 Insect Growth Regulators

Disrupt the molting, maturity from pupal stage to adult, or other life processes of insects.

1.2.23 Plant Growth Regulators

Substances (excluding fertilizers or other plant nutrients) that alter the expected growth, flowering, or reproduction rate of plants.

1.3 Pesticide History

Since the dawn of time, humanity has had two primary goals -obtaining enough food to survive and improving the quality of life. The single most important task facing the society is the production of food to feed its population before it can devote resources to education, arts, technology or recreation. A basic fact of green revolution in India which converted a food deficient state to food surplus state was the ability to control pests, weeds, insects and diseases by providing effective plant protection umbrella coupled with soil-water management and high yielding varieties. Today, farmers regard pesticides as an essential tool to ensure that they can maintain production of crops of quality and quantity to satisfy an increasing human population. Agricultural experts believe that these food and fiber needs can be met, but to do so will require the increased use of pesticides. Many people think that the use of pesticides started only a few decades ago. Actually mankind has a history of using crop protection products in the production of food and protection of environment. Going through the history, we see evolution of pest control products from non-selective, naturally occurring compounds to highly specific synthetic and biological materials that control specific pests. The following references are useful for understanding the history of the development of pesticides: Schrader (1942), Anonymous (1954), Schrader (1963), Kuhr (1971), Casida (1973), Fest and Schmidt (1973), Eto (1974), Kuhr and Dorough (1976), Matsmura (1985), Anonymous (1999), Handa (1999), Dunne (2002), Anonymous (2005), and Tomizawa and Casida (2005).

A brief review of development of plant protection products/approaches is given below:

1.3.1: Early Use Pattern up to 1850 AD (Era of Natural Products)

1200 BC	-	Use of salt and ash as non-selective herbicides by Biblical armies on the fields of the conquered
1000 BC	-	Homer refers to sulphur used in fumigation
100 BC	-	Romans used hellebore to control rats, mice and insects
25 BC	-	Virgil reports seed treatments using "niter and amurea"
900 AD	-	Chinese used arsenic to control garden insects
1300	-	Marco Polo writes about the use of mineral oil against mange on camels
1649	-	Rotenone used to paralyze fish in South America
1669	-	First mentions of the use of arsenic in western societies, used with honey in bait
1690	-	Tobacco extracts used as contact insecticides

1773 - Nicotine used as fumigant
1787 - Soaps
1787 - Terpentine emulsion as insecticide
1800 - Pyrethrums used in Caucasus
1800 - Spray of lime and sulfur recommended in insect control
- Whale oil used as scalecide
1810 - Arsenic dip recommended in insect-control
1820 - Sulphur recommended as a fungicide for mildew control in England
1822 - Mercuric chloride as insecticide
1845 - Phosphorus paste declared as official rodenticide in Prussia, also used for cockroach control
1848 - Rotenone derived from the roots of derris plant used as insect control in Asia

1.3.2: Era of Fumigants, Inorganics and Petroleum Products: 1850-1940

1854 - Carbon bisulfide (CS_2) used as fumigant in France
1860 - Mercuric chloride solutions used as fumigant to control soil born pests
1867 - Paris green (arsenical insecticide) used for Colorado beetle
1868 - Petroleum products used as insecticide in U.S.A.
1873 - Kerosene emulsions used as dormant sprays on deciduous trees
1877 - Hydrogen cyanide (HCN) used on firs as fumigant in museums
1880 - Lime sulphur used in U.S.A. against San Jose scale
1883 - Millardet discovered the use of Bordeaux mixture in France
1885 - Lead arsenate prepared and used against gypsy moth in Massachussets, U.S.A.
1886 - Pine resin used for scales
1892 - Dinitrophenol used as insecticide in Germany
1903 - First arsenic tolerance established in Britain
1911 - Carbon tetrachloride recommended as substitute for carbon disulfide in fumigation of grains
1920 - Chloropicrin used as fumigant in France
1929 - Anabasin isolated from plants and synthesized
1930's - Pentachlorophenol introduced as wood preservative against fungi and termites
1930's - *Bacillus thuringiensis* used as microbial insecticide
1932 - Methyl bromide (France) and thiocyanates as insecticides
1936 - Pure pyrethrum extract prepared
1938 - First synergist chemical discovered and sesame oil found to enhance effectiveness of pyrethrums

1.3.3 Era of Modern Synthetic Pesticides

Testing of synthetic chemicals for possible insecticidal value started during developmental programme for control of insect-born diseases among armed forces during Second World War. By 1947 more than 13000 such chemicals had been tested and classified. It was the development of synthetic organic pesticides during and post Second World War period that revolutionized use of chemicals for pest control on agricultural crops and pests of public health importance.

1.4 Insecticides

1.4.1: Organochlorines (OCs)

1939	-	DDT's insecticidal properties discovered in Switzerland by Paul Muller
1941, 42	-	BHC's insecticidal properties discovered in France and U.K.
1945	-	Chlordane synthesized
1947	-	Toxaphene synthesized
1948	-	Aldrin, dieldrin synthesized by Julius Hymen, U.S.A.
1948	-	Methoxychlor synthesized
*1949	-	DDT residues detected in Cow milk
1949	-	Heptachlor synthesized in U.S.A. by Julius Hymen, U.S.A.
1951	-	Endrin synthesized
1954	-	Mirex introduced
1955	-	Dicofol introduced
1956	-	Endosulfan synthesized
1957	-	Telodrin synthesized
1958	-	Chlordecone (Kepone) introduced
*1962	-	The book "Silent Spring" by Rachel Carson (1962) attracts international attention towards ill effects of pesticidal use
*1968	-	Insecticide Act passed in India to ensure safety from pesticides
*1970	-	Trials against DDT appear in U.S.A. and Sweden

* Some important events associated with the use of OCs

Because of high persistent nature coupled with high mammalian toxicity (in case of endrin, aldrin and dieldrin), most of the OCs have been banned globally.

1.4.2 Carbamates

Biologically active carbamates have been used as far back as the 17th century in the old Calabar region of southeast Nigeria.

1925-1935	-	Molecular structure of alkaloid 'Eserine' obtained from Calabar beans of a plant later named *Physostigma venenosum* was established is 1925 and compound renamed as physostigmine was synthesized in 1935

1947	-	Modern carbamate insecticides (Dimetan, Pyrolan, and Isolan) synthesized by Geigy, Switzerland
1956	-	Carbaryl (Sevin) by Union Carbide U.S.A.
1959	-	Propoxur (Baygon) introduced
1961	-	Mexacarbate (Zectran)
1962	-	Aminocarb, Mathiocarb
1963	-	Metacil
1965	-	Promecarb (carbamult)
1966	-	Aldicarb (Temik) and aldoxicarb (Aldicarb sulfone) introduced as first soil insecticide
1967	-	Carbofuran (Furadan)
1967	-	Methomyl (Lannate)
1968	-	Landrin, Dioxacarb
1969	-	Primicarb
1971	-	Bendiocarb (Ficam)

More recently introduced carbamates

2000- Indoxacarb, Alanycarb, -Furathiocarb, Carbosulfan (Advantage) and Fenoxycarb (Logic)

Because of appearance of pyrethroid insecticides in 1970's, onward, the development of carbamate slowed down.

1.4.3 Organophosphates (OPs)

Organophosphate is the most versatile chemical group, which has been explored to synthesize insecticides, fungicides, herbicides, nematicides besides chemosterilants. As OP insecticides were highly effective, thus required in relatively less amounts per unit area, and did not suffer from high persistence, therefore these toxicants rapidly replaced the use of OCs. Development of OPs was very rapid up to 1970's; however due to wide spread resistance development in insects to OPs and carbamates, their progress slowed down. Rather pyrethroids took their place.

Organophosphates are divided into three main groups

1.4.3.1 Aliphatic OPs

These are carbon chain like structures. The first OP used in Agriculture, TEPP 1943/46 (G. Schrader)* belonged to this group. Chronological development of OPs is given below:

1941	-	Schradan (Synthesized by G. Schrader)
1946	-	TEPP (Synthesis of TEPP improved by G. Schrader)
1949	-	Dimefox
1950	-	Malathion

1951 - Demeton
1953 - Mevinphos (Phosdrin)
1954 - Phorate
1955 - Dichlorvos (Vapona)
1956 - Ethion
1956 - Phosphamidon (Dimecron), Dimethoate (Rogor), Disulfoton (Disyoton), Ethion
1960 - Oxydemeton methyl (Metasystox)
1963 - Formothion , Dicrotophos (Bidrin), Dimethoate
1965 - Monocrotophos (Azodrin), Omethoate
1969 - Methamidophos (Monitor)
1971 - Acephete (Orthene)
1994 - Terbuphos

1.4.3.2 Phenyl OPs

The phenyl OPs contain a phenyl ring with one of the ring hydrogens replaced by attachment to the phosphorus moiety and other hydrogens replaced by Cl, NO_2, CH_3, CN or S. The phenyl OPs are generally more stable than the aliphatic, thus their residues are longer lasting. The first phenyl OP brought to agricultural use was ethyl parathion in 1947.

1947 - Parathion – Synthesized by G. Schrader in 1944
1949 - Methyl parathion
- Profenophos (Curacron)
- Sulprofos (Bolstar)
- Isofenphos (Oftanol)
1950 - EPN
1952 - Dichlorophos, Chlorthion, Diazinon, Trichlorfon (Diptrex)
1954 - Dicapthon , Ronnel
1957 - Fenitrothion (Sumithion)
1958 - Fenthion (Dasanit)
1961 - Chlorfenvinphos
1961/
1964 - Phenthoate
1962 - Crotoxyphos (Lebaycid)
1965 - Leptophos (Phosvel)
1966 - Tetrachlorvinphos (Gardona), Iodofenphos
1973 - Fenamiphos (Nemacur)

1.4.3.3 Heterocyclic OPs

The OPs of this group contain heterocyclic ring in their structure. The first compound of this group was diazinon introduced in 1952 followed by other toxicants as under:

1951 - Coumaphos (Co-Ral)
1952 - Diazinon

1953 - Azinphos-methyl (Guthion), Azinphos-ethyl (Guthion)
1956 - Morphothion (Ekatin M)
1961 - Phosalone (Zolone)
1963 - Methidathion (Supracide)
1965 - Chlorpyriphos (Dursban), Quinalphos
1966 - Phosmet (Imidan)
- Isazophos (Brace, Triumph)
- Chlorpyriphos ethyl (Reldan)
1967 - Triazophos (Hostathion)
1970 - Primiphos methyl
1972 - Primiphos ethyl

* G. Schrader is called father of organophosphate insecticides because of his significant discovery of OPs.

1.4.4 Pyrethroids

Natural pyrethrum (pyrethrins) has seldom been used for agricultural purposes because of its cost and instability in sunlight. In recent decades, many synthetic pyrethrins like materials have become available. They were originally referred to as synthetic pyrethroids. Currently the better nomenclature is simply pyrethroids. These are stable in sunlight and are generally effective against most agricultural insect-pests when used at very low rates. Pyrethroids consist of four generations.

1.4.4.1 First Generation Pyrethroids

More stable and of longer residual activity than natural pyrethrins

1949 - Allethrin introduced, (synthesized in 1945)

1.4.4.2 Second Generation Pyrethroids

These are also more stable than natural pyrethrins, but they too decompose rapidly on exposure to air and sunlight, hence not used in agriculture.

1961 - Dimethrin
1965 - Tetramethrin
1967 - Resmethrin
1969 - Bioresmethrin
1969 - Bioallethrin
1973 - Phenothrin (Sumithrin)

1.4.4.3 Third Generation Pyrethroids

These are virtually unaffected by ultraviolet in sunlight lasting 4-7 days as efficacious residues on crop foliage, hence they achieved wide application in agriculture.

1976 - Fenvalerate
1977 - Permethrin, Cypermethrin (Synthesized in 1972-73)
1977 - Deltamethrin
1980 - Fenpropathrin (Acaricide) (Synthesized in 1971)

1.4.4.4 Fourth Generation Pyrethroids (1975-1983)

These are photo stable i.e. they do not undergo photolysis in sunlight. Because of minimal volatility, they provide effective residue persistence up to 8-10 days under optimum conditions.

1980 - Fenpropathrin
1981 - Cyfluthrin
1983 - Fluvalinate
1983 - Alphamethrin
1986 - Esfenvalerate
1986 - Ethofenoprox
1986 - Telomethrin
1987 - Tefluthrin, Tralomethrin
1997 - Bifenthrin
1998 - Imiprothrin, Acrinathrin
2002 - Lambda cyhalothrin

1.4.4.5 Non-Ester Pyrethroids

These are photostable, and possess wide spectrum activity to insects-pests, have low mammalian and fish toxicity. These have much higher lipophilic properties than esters.

- Etofenprox
- Flufenprox

1.4.5 Neonicotinoids

The last decade of twentieth century was introduction of neonicotinoids, the newest major class of insecticides having outstanding potency and systemic action for crop protection against piercing-sucking pests and soil insects. They are highly effective for flea control on cats and dogs. Some of these are given below:

1990's - Imidacloprid (in Europe and Japan)
2002 - Acetamiprid
2002 - Thiacloprid
2002 - Thiamethoxam
2003 Clothianidin
Dinotefuran
Nitenpyram

1.4.6 Fiproles (Phenylpyrazoles)

1990 - Fipronil (Regent ®, Frontline ® Introduced/ Registered in USA in 1996 for crop use. It is effective against insects resistant to pyrethroids, OPs and carbamate insecticides).

1.4.7 Benzoyl Ureas

Benzoyl ureas are an entirely different class of insecticides that act as insect growth regulators (IGR). They interfere with chitin synthesis and are taken by ingestion.

1978-80 - Triflumuron, Chlorfluazuron, Teflubenzuron, Hexaflumuron and Flufenoxuron

1980 - Flucycloxuron, Flurazuron, Novaluron, Diafenthiuron, Bistrifluron and Noviflumuron

1982 - Diflubenzuron (Dimlin) U.S.A.

1990 - Lufenuron

1993 - Hexaflumuron

2001 - Novaluron

1.5 Fungicides

2001 - Novaluron

1940's - First dithiocarbamate fungicide, zineb registered for use

- First appearance of dicarboximide fungicide captan

1950's - Introduction of maneb fungicide

- First use of streptomycin for fungal and bacterial disease control

1960's - Introduction of mancozeb broad spectrum fungicide

- First systemic fungicide carboxin introduced
- Introduction of systemic fungicide family with benomyl registered

1963 - Kitazin (OP) introduced

1970's - Introduction of Ficam. Most use patterns for mercury based compounds cancelled

1980's - Introduction of matalaxyl systemic fungicide

- Controversial removal of daminozide from apple market

1980's - The close of the twentieth century brought evolution of newer classes of highly specific, low toxicity and low use rate insecticides and fungicides that featured both forward and backward systemic activity to control diseases like apple scab in apple

1990's & 2000 - Introduction of low use rate sterol inhibitor fungicides including **myclobutanil** and later **flusilazole** which provided both protectant and eradicant disease control for apples and grapes

- Introduction of new generation systemic fungicides including **propamacarb**, **dimethomorph** and **cymoxanil**
- Introduction of first *strobilurin* class (azoxystrobin) of broad spectrum and systemic fungicide based on natural occurring fungicide for use on apple and potatoes

1.6 Herbicides

1.6.1: Inorganic Compounds

Biblical Times - Brine and mixture of salt and ashes used by Romans to sterilize conquered land

1896 - Copper sulfate used to kill weeds in grain fields

1906-1960 - Sodium arsenite solutions were standard herbicide in commerce

1942 - Ammonium sulfamate used for bush control

- Borate salts used heavily for soil sterilization

- Sodium chlorate dominated as soil sterilant for about 50 yrs

EPA, U.S.A. placed heavy restrictions on some inorganics because of their persistence in soil

1.6.2: Organic Compounds

1940 - Introduction of 2,4-D, the first growth regulating herbicide

1943 - MCPA introduced

1952 - Triazine herbicides, Atrazine

1959 - Bypridylium herbicide **paraquat** which provided fast activity and desiccation properties

1960 - Treflan (trifluralin) soil applied herbicide for broad spectrum grass and broad leaf weeds

1963 - Diphenyl ether (Nitrofen), pre-emergent and early post emergent herbicide

1970's - Pre-plant herbicides like **triallate** to control wild oats in cereal crops

1972 - Glyphosate herbicide introduced

1978 - Bentazon for post-emergent broad leaf weed control in soybean and corn

1979 Benzamide class herbicide **alachlor** for pre-emergent weed control in corn and soybean

1981 - Introduction of **bromoxynil** for post emergent broad leaf control in cereals and corn

1982 - Dicamba for broad leaf weed control

1983 - Metribuzin for pre-emergent weed control in soybean and pulse crops

1989 - Diclofop- methyl for post emergent use to control grassy weeds including wild oats in cereal crops
1988 - Metolachlor as pre-emergent herbicide in soybean and corn
1990 - Triazolopyrimidines (Flumetsulam) introduced
1991 - Introduction of several low use rate* Group-I (ACC ase control) and Group-2 (ALS/AHAS inhibitors) graminicides including imazethapyr, nicosulfuron, fenoxaprop-ethyl, tralkoxydim and clodinafop propargyl, metsulfuron-methyl and tribenuron-methyl for broad leaf weeds in cereals
1995 - Introduction of clopyralid for thistle and broad leaf weed control in cereals and corn
1995 - Development of herbicide tolerant crops (Round-up), Ready (glyphosate), Liberty Link (glufosinate) and Smart (imazethapyr,) and related molecules. Crops bioengineered to sustain broad spectrum herbicides include cotton, corn, canola, soybeans, rice, sugar beet, wheat and turf
1996 - Introduction of glufosinate ammonium broad spectrum non-selective herbicide based on a soil bacterium
1998 - Introduction of imazapyr (Arsenal)
1999-2000 - Introduction of new novel herbicide classes
2001 - (i) Benzoyl cyclohexanedione – Mesotrione (Callisto registered with EPA
2003 - (ii) Pyrimidin diones – Butafenacil (Inspired) registered in U.S.A.
2003 - Unclassified (Miscellaneous) – Flufenpyr-ethyl
2004 - Triazolinones – Amicarbazone – (Dinamic^{R})

*Introduction of low dose herbicides and herbicide tolerant varieties of crops have reduced use of herbicides by several thousand tons over a four year period

1.7 Nematicides

Most of the early used nematicides have been fumigants. It is only after 1950's that some organophosphate and carbamate compounds have been introduced as nematicides. Recently some nematicides of natural and microbial origin have been introduced in U.S.A. Year wise development of nematicides is given below:

1869 - Carbon disulfide
1936 - Chloropicrin
1940's - Methyl bromide
-do- - DD (1,3-D+1,2-D) Dichloropropene-Dichloropropane
-do- - DBCP
-do- - Telone II (1,3-D)
-do- - Formaldehyde

1950's	-	Metham Sodium (Vapum)
-do-	-	Furadan (Carbofuran)
1960's	-	Mocap (Ethoprop)
1965	-	Fensulfothion (Dasanit)
1966	-	Thionazin (Nemafos)
1966	-	Temik (Aldicarb)
1970's	-	Vydate (Oxamyl)
1973	-	Fenamiphos (Nemacur)
1980's	-	Enzone (Gy-81) Sodium tetrathiocarbonate
-do-	-	Nematrol (Sesamechaff)
1990's	-	Ditera (Microbial)

1.8 Miscellaneous Classes of Softer Pesticides/ Approaches

1940	-	**Synergists** for pyrethrum i.e. piperonyl butoxide
1942	-	**Insect repellents, Dimethyl phthalate, Benzoyl benzoate** and other insect-repellents were discovered and developed for military use and subsequently for civilian use
1999	-	**Repellant** for cockroaches and ants, N-methyl-neodecanamide was developed
1930/1960	-	**Microbial insecticides, *Bacillus thuringiensis*** (Bt) though developed in 1930; it was registered in 1960 for use on lettuce and cole crops
1967	-	Chemosterillants, Insect-growth regulators (IGR) and Hormone mimics developed in U.S.A
1970	-	Integrated Pest Management (IPM) concept introduced
1978-80	-	Pheromones introduced in U.S.A on cotton to manage pink boll worm
1980's	-	Introduction of plant pesticides/transgenic pesticides/genetically modified crops like corn resistant to corn borer and cotton resistant to *heliothis*
1980	-	Avermectins introduced in U.S.A.
1995	-	Bt cotton registered
1996	-	Avermectin B1a
1997	-	Spinosad (spinos AD) introduced
1990-2005	-	Herbicide resistant crops like (Round up), Ready for soybean, corn etc.

Part B

1.9 Principal Pesticides

The availability of a wider range of chemicals and international marketing of pesticides has led to a global growth in the use. The following sections provide a brief account of some of the pesticides now available.

1.9.1 Insecticides

Initially the two main chemical groups of insecticides i.e. organochlorine (OCs) and organophosphates (OPs), were available and both were neurotoxic in action. The OC insecticides, including DDT, dieldrin and endrin, had one main advantage, namely their persistence that enabled farmers to achieve control over a long period of time. Later it was realised that this attribute led to their residues remaining in the environment and accumulated in some animals at the end of food chain. In consequence, residues of these chemicals can be found everywhere, although their use on agricultural crops has been banned since long, except for endosulfan. However, DDT, is still in use on a limited scale in vector control only.

Organophosphate insecticides are a diverse group (Anonymous 1999), some of which are extremely toxic (e.g. parathion, methidathion and monocrotophos), while others, such as temephos, malathion and trichlorfon, are much less hazardous to use. When used in place of the OC insecticides, more people suffered acute poisoning, as the need for protective clothing had not been adequately recognised in many countries. Karalliedde et al., (2001) provide a critical overview of organophosphates and their impact on health.

Another group with a similar mode of action is the carbamates, though these also vary very much in their toxicity. The most toxic examples are aldicarb and carbofuran. The less-toxic carbaryl has been very widely used as a broad-spectrum insecticide. Newer insecticide groups include the pyrethroids and neonicotinoids.

Pyrethroids (cypermethrin, cyfluthrin, deltamethrin, fenvalerate etc.) helped in overcoming against OP compounds. These were highly effective against a variety of insect-pests at very low doses.

More recently, the neonicotinoids, notably imidacloprid, have been rapidly accepted, especially where insects are resistant to earlier types of insecticides. Most salient feature of these newer chemicals viz, imidacloprid and fipronil, a phenylpyrazole, is that they are active at extremely low dosages.

1.9.2 Herbicides

Herbicides are the most extensively used group of pesticides, except by small scale farmers in Third World countries. They can act on contact with a plant or are translocated within the plant. Good spray coverage is needed with contact herbicides. Sometimes, only part of the foliage is affected by some weeds, although adversely affected, will survive. A few important groups are mentioned here. Most have a very low mammalian toxicity. Most concern regarding human toxicity has been directed at paraquat, as it is lethal if the concentrated liquid reaches the lungs. A detailed assessment of paraquat poisoning has been reported by Lock and Wilks (2001).

1.9.2.1 Aryloxyphenoxy Propionates

These have good activity against grass weeds in broad-leaf crops as a post emergence translocated herbicide. One example is fluazifop-butyl.

1.9.2.2 Bipyridyliums

Paraquat is the most important in this group. It damages foliage quickly on contact, but is also very strongly adsorbed on to the soil and rendered ineffective. The rapid wilting and desiccation of foliage within hours has enabled effective weed control to be achieved in many crops, where the spray is directed away from the actual crop. It has been extensively used in tree crops, such as rubber plantations.

1.9.2.3 Dinitroanilines

Trifluralin is a good example of a pre-planting soil-incorporated herbicide to reduce the impact of grass weeds in a broad-leaf crop. Low water solubility minimizes leaching and movement within the soil, but being volatile they must be covered by the soil.

1.9.2.4 Phenoxy or 'Hormone' Herbicides

2, 4-D and MCPA are highly selective for broad-leaf weeds, being translocated throughout the plant, affecting cellular division.

1.9.2.5 Phosphono Amino Acids

Glyphosate and glufosinate are foliar-applied, translocated herbicides that interfere with normal plant amino acid synthesis. They are non-selective, but more effective against grasses than broad-leaf weeds. There is no soil activity. They are formulated to improve uptake by the plants as rainfall shortly after application can reduce effectiveness.

1.9.2.6 Substituted Ureas

Most of these, such as isoproturon, flumeturon, diuron and linuron, are non-selective, pre-emergence herbicides, which are absorbed in the soil and then taken up by roots. Some are active as foliar-applied, post-emergence herbicides.

1.9.2.7 Sulfonylureas

This is a large group that is used mainly to control broad-leaf weeds by inhibiting meristematic growth. Metsulfuron- methyl and others in the group have both foliar and soil activity and are active at extremely low application rates - a few grams per hectare. However, if small amounts remain active in the soil too long, the following crop may be adversely affected.

1.9.2.8 Triazines

This group includes one of the most used herbicides, atrazine, which was very effective as a post-emergence spray in maize. However, it has been implicated in environmental problems. It has been observed that very low doses of atrazine in water have an endocrine disruption effect that has resulted in decline in frog populations. Therefore, it has been withdrawn from certain uses.

1.9.3 Fungicides

Mostly, developments of fungicides have occurred only in the past few decades. Apart from the contact, protectant fungicides, such as copper fungicides and mancozeb, a number of systemic fungicides with different modes of action have been developed (Hewitt, 1998), most recently the strobilurins. List of some important groups of fungicides is given in the Table 1.1.

Table 1.1: Some examples of fungicides

Type of fungicide	Example
Triazloes	propiconazole, tebuconazole
Morpholines	fenpropimorph
Anilinopyrimidines	cyprodinil
Benzimidazoles	carbendazim
Carboxamides	carboxin (only in mixtures)
Strobilurins	azoxystrobin
Others	chlorothalonil

1.9.4 Rodenticides

Various rodenticides have been used in baits, usually inside traps to prevent other mammals (especially dogs) from gaining access to the poison. Following the use of the anti-coagulant warfarin, to which rats have become resistant, other rodenticides such as bromadiolone and difenacoum have been introduced. There is particular concern that predatory birds can be affected by eating rodents that have consumed a poisoned bait, but have not yet died.

Part C

1.10 Pesticide Usage Pattern

1.10.1 Background

India is a densely populated country with about 15% of the world population and 2.5% of the world geographical area. About 40% of the area is available for cultivation. India's population, at present, is about 1.6 billion and is predominantly an agricultural country. The total food grain production has risen from 50.82 million MT in 1950-51 to an estimated quantity of 229.85 millions tons during 2008-09. In order to meet the needs of a growing population, agricultural production and protection technology have to play a crucial role. Substantial food production is lost due to insect pests, plant pathogens, weeds, rodents, birds, nematodes and in storage.

1.10.2 Crop Losses Due to Pests

The losses of crops caused by pests and plant diseases are quite high both in developed and developing countries. During the last 3-4 decades, chemical control of pests aimed at minimizing these losses has been introduced throughout the world. Approximately 70% of the pesticides used in the world are applied in developed countries and 30% in the developing countries (Pimentel, 1987).According to Ware and Whitcare (2004), about one-third of the world food crops are destroyed by pests during growth, harvesting and storage. Losses are even higher in developing countries.

It is imperative to contain these colossal losses, year after year, in order to meet the rising demand for food grains and other agricultural commodities. The concept of food safety is changing considerably. Safe food is not only expected to look fresh and tasty but it must be free from all the deleterious contaminants. The presence of pesticide residues in food products is a clear violation of this simple reasonable expectation (Singh and Dhaliwal, 2000). Hence, this has to be done in such a way that environmental quality is maintained and long-term sustainability of the agro ecosystem does not suffer.

Pesticides industry has developed substantially and it contributes significantly towards India's agriculture and public health. In India, crops are affected by over 200 major pests, 100 plant diseases, hundreds of weeds and other pests like nematodes, harmful birds and rodents. About 4,800 million rats cause havoc in India. Approximately, 30% of Indian crop yield potential is being lost due to insects, disease and weeds which in terms of quantity would mean 30 million tones of food grains. It is estimated that up to 35% of the total food production is lost, due to different types of pests, at different stages of crop growth and during storage. The value of total loss has been placed at Rs. 90,000 crores in 2002 and in 2007, it is staggering Rs. 1,40,000 crores (Anonymous, 2008), represents about 18% of the gross national agriculture production. The pest wise losses are shown in Figure 1.1.

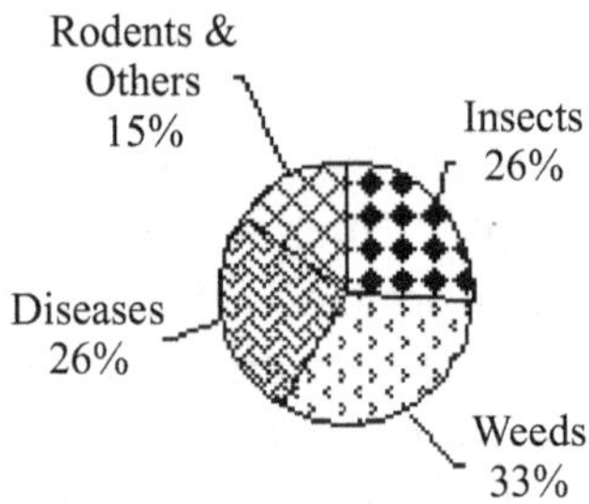

Figure 1.1: Losses caused by different pests (Source: Anonymous, 2008)

Avoidable crop losses due to pests and cost benefit ratio of pesticides use in different crops have been shown in Table 1.2.

Table1.2: Avoidable crop losses due to pests & cost benefit ratio of pesticides use in different crops

Crop	Avoidable Losses (%)	Cost:Benefit*
Cotton	40-90	1:7
Paddy	21-51	1:7
Mustard	35-75	1:12
Sunflower	36-51	1:8
Groundnut	29-42	1:26
Maize	20-25	1:3
Pulses	40-88	1:4
Sugarcane	8-23	1:13
Vegetables	30-60	1:7
Fruits	20-35	1:4

(*For instance, Cost:Benefit of 1:8 means that for every one rupee spent on pesticides, farmer gets a benefit of value Rs. 8) (Source: Anonymous, 2008)

Today, the domestic industry represented is characterized by over-capacity, low capacity utilization and unsustainable levels of production from many units and low investments in R&D. Besides, the formulation market is highly fragmented with large number of small formulators engaged in the production of formulations for traditional technical pesticides largely aiming to meet the requirement of the domestic market. The domestic market is less than 2 % of the global agrochemical market and is largely confined to agrochemicals for a limited number of crops like cotton, wheat and rice. Globally, there is a growing trend towards low dosage, high potency molecules and as such, market for usage of high volume pesticides is declining.

1.10.3 Global Scenario

Global generic market is Rs. 45000 crores (US$ 10 billion). Out of this, the opportunity for Indian companies is immense; for instance 10% of US$ 20 billion would equal US$ 2 billion, given a favourable climate. We have already achieved US$ 500 million milestone.

1.10.4 Indian Scenario

1) India is the 4th largest producer of Agrochemical after U.S.A., Japan and China.
2) India is the second largest producer of pesticides in Asia.
3) India exports pesticides for approximately Rs. 2800 crores.
4) Total agrochemical market in India = Rs. 4500 crores.

The Indian pesticides industry can be broadly divided into three categories, Multi-National Companies (MNCs), Indian companies including the Public Sector companies and Small Scale Sector Units. Besides about 60 Indian companies in the organized sector manufacturing pesticides, there are around 10 multi-national companies operating in the country. Most Indian technical manufacturers are focused on off-patent pesticides, which comprise over 70% of the Indian market.

1.10.5 Production of Pesticides

The indigenous capacity in pesticides sector is adequate to meet the domestic demand. The production data for the last six years is as under:

Table 1.3: Production of pesticides

Year	Production (in tonnes)
2001-02	81803
2002-03	69565
2003-04	85118
2004-05	93966
2005-06	82240
2006-07	84701

(Source: Anonymous, 2008a)

1.10.6 Demand of Pesticides

The present demand of various types of pesticides for use in agriculture during 2007-08 was 42017 MT (technical grade). India is able to meet 95% of its demand of pesticides domestically. The demand of pesticides is worked out by the State / Union Territories (UTs) keeping in view several factors like crop production programmes, targeted area proposed to be brought under plant protection coverage, past consumption, inter-substitutability for a given pesticide, package of practices recommended by State Agricultural Universities (SAUs), Indian Council of Agricultural Research Institution (ICAR) etc. These are discussed and finalized during the Zonal/National Conference on inputs for *Kharif* and *Rabi* crops, organized by the Ministry of Agriculture. The total demand of pesticides for the 2002-03 to 2007-08 and state wise demand of pesticides is shown in Tables 1.4 and 1.5, respectively.

Table1.4: Demand of pesticides

Year	Demand in MT (Tech. Grade)
2002-03	49220
2003-04	48736
2004-05	45412
2005-06	44324
2006-07	43718
2007-08	42017

(Source: Anonymous, 2008a)

Table 1.5: State wise demand of pesticides for the years 2003-04 to 2007-08

Sr. No.	States/UTs	2003-04	2004-05	2005-06	2006-07	2007-08
1.	Andhra Pradesh	3600	2045	2045	1500	1300
2.	Assam	215	205	195	190	186
3.	Arunachal Pradesh	17	17	17	17	4
4.	Bihar	975	930	925	935	960
5.	Chhattisgarh	450	704	535	535	570
6.	Goa	5	5	5	80	7
7.	Gujarat	4200	4550	3400	3000	2750
8.	Haryana	4732	4700	4500	4650	4550
9.	Himachal Pradesh	470	332	340	297	288
10.	Jammu & Kashmir	110	101	28	1154	26
11.	Jharkhand	180	60	55	85	72
12.	Karnataka	2600	2400	2400	2200	1975
13.	Kerala	1690	346	454	368	364
14.	Madhya Pradesh	938	921	880	939	961
15.	Maharashtra	3500	3300	3200	3188	3200
16.	Manipur	27	29	32	33	33
17.	Meghalaya	7	7	9	100	10. 328
18.	Mizoram	20	20	46	61	55
19.	Nagaland	7	7	7	7	7
20.	Orissa	1131	1131	1131	1131	1155

21.	Punjab	7100	7000	7050	6450	6450
22.	Rajasthan	3175	3150	3250	3250	3050
23.	Sikkim	4	4	NIL	NIL	1
24.	Tamil Nadu	2564	2619	2634	2369	2182
25.	Tripura	18	18	28	25	17
26.	Uttar Pradesh	6750	6725	6700	6675	7075
27.	Uttaranchal	131	179	147	150	259
28.	West Bengal	4000	3800	4200	4400	4400
29.	A & N Islands	3	9	9	9	19
30.	Chandigarh	1	0.75	0.75	0.75	N/A
31.	Delhi	55	50	50	33	50
32.	Dadra & Nagar Haveli	6	6	6	6	N/A
33.	Daman & Diu	1	1	1	1	N/A
34.	Lakshadweep	2	2	2	2	N/A
35.	Pondicherry	52	38	42	42	41
	Total:	**48,736**	**45,412**	**44,324**	**43,718**	**42017**

N/A: Not available.

(Source: Anonymous, 2008a)

1.10.7 Consumption of Pesticides

The consumption of technical grade of pesticides in the states of India during 2001-02 to 2006-07 is as under:-

Table 1.6: Consumption of pesticides

Year	Consumption in MT (Tech. Grade)
2001-02	47020
2002-03	48350
2003-04	41020
2004-05	40672
2005-06	39773
2006-07	37959

(Source: Anonymous, 2008a)

Crops, like cotton, wheat and rice together account for 70% of the total agrochemical consumption. On the other hand, global agrochemical consumption on the commercial crop basis is dominated by fruits and vegetables. The state of U.P., Punjab, Haryana, West Bengal and Maharashtra account for around 62% of the total market. The reasons for significantly lower usage of herbicides and fungicides in India are two fold. First, weeding in India is done manually and second, the tropical climate is more conducive for the growth of insects as compared to herbs/fungi.

1.10.8 Consumption Pattern in India as Compared to Other Countries

The consumption of pesticides in India is low in comparison to other countries. The consumption pattern of pesticides as available from Agrochemicals Sub-Group of Indian Chemical Council is as under:-

Table 1.7: Per hectare consumption of pesticides

Country	Consumption (Kg/ha)
Korea	16.56
Japan	10.80
China	2.0-2.5
Europe	1.90
USA	1.50
Thailand	1.37
Indonesia	0.58
India	0.38

(Source: Anonymous, 2002a)

There is a marked difference in the consumption pattern of agrochemicals in India vis-à-vis the rest of the world. Insecticides account for 76% of the total domestic market. On the other hand, herbicides and fungicides have a significantly higher share in the global market. Consumption of pesticides from 2000-2007 is given in Table 1.8.

Table 1.8: Consumption of pesticides (M.T. Technical Grade) in various states during 2000-01 to 2006-07

Sr. No.	States	2000-1	2002-3	2003-4	2004-5	2005-6	2006-7
1.	Andhra Pradesh	4000	3706	2034	2133	1997	1051
2.	Assam	245	181	175	170	165	165
3.	Arunachal Pradesh	13	15	147	17	2	17.10
4.	Bihar	853	1010	860	850	875	890
5.	Chhattisgarh	NA	NA	332	486	450	490
6.	Gujarat	2822	4500	4000	2900	2700	2670
7.	Goa	6	5	5	5	5	6.9
8.	Haryana	5025	5012	4730	4520	4560	4600
9.	Himachal Pradesh	302	380	360	310	300	290
10.	Jammu & Kashmir	1	98	9	12	1433	829
11.	Jharkhand	150	40	56	69	70	57

12.	Karnataka	2020	2700	1692	2200	1638	620
13.	Kerala	754	902	326	360	571	547
14.	Madhya Pradesh	871	1026	662	749	787	879
15.	Maharashtra	3239	3724	3385	3030	3198	3193
16.	Manipur	20	19	25	26	28	27
17.	Meghalaya	6	6	6	8	6	9.3
18.	Mizoram	8	15	15	25	25	40
19.	Nagaland	8	7	7	5	5	5
20.	Orissa	1006	1134	682	692	963	1068
21.	Punjab	7005	7200	6780	6900	5610	5710
22.	Rajasthan	3040	3200	2303	1628	1008	1523.20
23.	Sikkim	4	3	3	No cons. as orga-nic State	No cons. as organic State	2
24.	Tamil Nadu	1668	3346	1434	2466	2211	2048
25.	Tripura	11	88	118	17	14	9
26.	Uttar Pradesh	7023	6775	6710	6855	6672	7022
27.	Uttaranchal	99	129	147	132	141	139
28.	West Bengal	3250	3000	3900	4000	4250	3950
29.	Andaman & Nicobar	3	3	6	3	3	4.415
30.	Chandigarh	3	1	0.78	0.78	0.78	NA
31.	Delhi	55	60	56	53	39	57
32.	Dadra & Nagar Haveli	6	5	5	5	4	NA
33.	Daman & Diu	2	1	1	1	1	NA
34.	Lakshadweep	2	2	2	2	1	NA
35.	Pondicherry	65	57	46	42	41	40.46
	Total (in round figure)	**43584**	**48350**	**41020**	**40672**	**39773**	**37959**

NA: Not available.

(Source: Anonymous, 2005a)

1.10.9 Export and Import of Pesticides

India is a net exporter of agrochemicals. The key export destination markets are USA, UK, France, Netherlands, Belgium, Spain, South Africa, Bangladesh, Malaysia and Singapore. Some of the agro-chemicals exported over the years include cypermethrin, isoproturon, endosulfan and aluminium phosphide. Exports consist mostly of off-patent products. The value of export and import during 2005-06 were Rs.2790.69 and Rs.754.41 crores, respectively.

Table 1.9: Insecticides registered under/section 9 (3) of the Insecticide Act, 1968 as on 13/11/2009

Sr. No.	Name of the Pesticide
1.	2,4-Dichlorophenoxy Acetic Acid**(H)**
2.	Acephate **(I)**
3.	Acetamiprid**(I)**
4.	Alachlor **(H)**
5.	Allethrin **(I)**
6.	Alphacypermethrin **(I)**
7.	Alphanaphthyl Acetic Acid **(PGR)**
8.	Aluminium Phosphide**(R, I, Fg)**
9.	Anilophos**(H)**
10.	Atrazine**(H)**
11.	Aureofungin**(F)**
12.	Azadirachtin (Neem Products) **(Bio)**
13.	Azoxystrobin**(F)**
14.	*Bacillus thuringiensis* (B.t.) **(Bio)**
15.	*Bacillus thuringiensis* (B.S.) **(Bio)**
16.	Barium Carbonate **(Rat Poison)**
17.	*Beauveria bassiana***(B)**
18.	Bendiocarb**(I)**
19.	Benfuracarb**(I)**
20.	Benomyl**(F)**
21.	Bensulfuron **(H)**
22.	Beta Cyfluthrin**(I)**
23.	Bifenazate**(A)**
24.	Bifenthrin**(I)**
25.	Bispyribac Sodium**(H)**
26.	Bitertanol **(F)**
27.	Bromadiolone**(R)**
28.	Buprofezin **(I)**
29.	Butachlor**(H)**
30.	Captan **(F)**
31.	Carbaryl **(I, PGR)**
32.	Carbendazim **(F)**
33.	Carbofuran **(I)**
34.	Carbosulfan **(I)**
35.	Carboxin**(F)**
36.	Carfentazone Ethyl**(H)**
37.	Carpropamid **(F)**
38.	Cartap Hydrochloride**(I)**

39. Chlorantraniliprole **(I)**
40. Chlorofenvinphos**(I)**
41. Chlorfenapyr**(I)**
42. Chlorimuron ethyl**(H)**
43. Chlormequat Chloride (CCC) **(PGR)**
44. Chlorothalonil**(F)**
45. Chlorpyriphos **(I)**
46. Chlorpyriphos Methyl **(I)**
47. Cinmethylene **(H)**
48. Clodinafop-propargyl (Pyroxofop-propargyl)**(H)**
49. Clomazone **(H)**
50. Chlothiandin **(I)**
51. Copper Hydroxide **(F, N, M)**
52. Copper Oxychloride **(F)**
53. Copper Sulphate**(Al, F)**
54. Coumachlor **(R)**
55. Coumatetralyl **(R)**
56. Cuprous Oxide **(F)**
57. Cyfluthrin **(I)**
58. Cyhalofop-butyl **(H)**
59. Cymoxanil **(F)**
60. Cypermethrin **(I)**
61. Cyphenothrin **(I)**
62. Dazomet **(F, H, N)**
63. Deltamethrin (Decamethrin)**(I)**
64. Diazinon **(A)**
65. Dichloro Diphenyl Trichloroethane(DDT) **(I)**
66. Dichloropropene and Dichloropropane Mixure (DD-mixture**(N)**
67. Dichlorvos (DDVP) **(I)**
68. Diclofop-Methyl **(H)**
69. Dicofol **(A)**
70. Difenocenazole **(F)**
71. Difenthiuron **(A)**
72. Diflubenzuron **(I)**
73. Dimethoate **(I)**
74. Dimethomorph **(F)**
75. Dinocap **(F)**
76. Dithianon **(F)**
77. Diuron **(H)**
78. Dodine **(F)**
79. D-trans Allethrin **(I)**

80. Edifenphos **(F)**
81. Emamectin Benzoate **(I)**
82. Endosulfan **(I, A)**
83. Ethephon **(PGR)**
84. Ethion **(I,A)**
85. Ethofenprox (Etofenprox) **(I)**
86. Ethoxysulfuron **(H)**
87. Ethylene Dibromide and Carbon Tetrachloride mixture (EDCT mix3:1) **(I, N)**
88. Famoxadone **(F)**
89. Fenamidone **(F)**
90. Fenarimol **(F)**
91. Fenazaquin **(I,A)**
92. Fenitrothion **(I)**
93. Fenobucarb(BPMC) **(I)**
94. Fenoxaprop-p-Ethyl **(H)**
95. Fenpropathrin **(I,A)**
96. Fenpyroximate **(A)**
97. Fenthion **(I)**
98. Fenvalerate **(I,A)**
99. Fipronil **(I)**
100. Flubendimide **(I)**
101. Fluchloralin **(H)**
102. Flufenacet **(H)**
103. Flufenoxuron **(I)**
104. Flufenzine **(I)**
105. Flusilazole **(F)**
106. Fluvalinate **(I,A)**
107. Forchlorfenuron **(PGR)**
108. Fosetyl-AI **(F)**
109. Gibberellic Acid **(PGR)**
110. Glufosinate Ammonium **(H)**
111. Glyphosate **(H)**
112. Hexaconazole **(F)**
113. Hexazinone **(H)**
114. Hexythiazox **(A)**
115. Hydrogen Cyanamid **(I,R)**
116. Imazethapyr **(H)**
117. Imidacloprid **(I)**
118. Imiprothrin **(I)**
119. Indoxacarb**(I)**
120. Iprobenfos (Kitazin) **(F)**
121. Iprodione **(F)**

122. Isoprothiolane **(F)**
123. Isoproturon **(H)**
124. Kasugamycin **(F)**
125. Kresoxim Methyl **(F)**
126. Lambdacyhalothrin **(I,A)**
127. Lime Sulphur **(F, I, B)**
128. Lindane **(I)**
129. Linuron **(H)**
130. Lufenuron **(H)**
131. Magnesium Phosphide Plates **(F)**
132. Mancozeb **(F)**
133. Malathion **(I, A)**
134. Mepiquate Chloride **(PGR)**
135. Mesosulfuron Methyl + lodosulfuron methyl **(H)**
136. Metaflumizone **(I)**
137. Metalaxyl **(F)**
138. Metalaxyl-M **(F)**
139. Metaldehyde **(M)**
140. Methabenzthiazuron **(H)**
141. Methomyl **(I, A)**
142. Methoxy Ethyl Mercury Chloride (MEMC)
143. Methyl Bromide **(I, A, R)**
144. Methyl Chlorophenoxy Acetic Acid (MCPA) **(H, F)**
145. Methyl Parathion **(I)**
146. Metiram **(F)**
147. Metolachlor **(H)**
148. Metribuzin **(H)**
149. Metsulfuron Methyl **(H)**
150. Milbemectin **(A, I)**
151. Monocrotophos **(I, A)**
152. Myclobutanil **(F)**
153. Novaluron **(IGR)**
154. Nuclear polyhyderosis virus of *Helicoverpa armigera* **(Bio)**
155. Nuclear polyhyderosis virus of *Spodoptera litura* **(Bio)**
156. Oxadiargyl **(H)**
157. Oxadiazon **(H)**
158. Oxycarboxin **(F)**
159. Oxydemeton-Methyl **(I)**
160. Oxyfluorfen **(H)**
161. Paclobutrazole **(PGR)**
162. Paraquat dichloride **(H)**
163. Penconazole **(F)**
164. Pencycuron **(F)**

165. Pendimethalin **(H)**
166. Permethrin **(I)**
167. Phenthoate **(I,A)**
168. Phorate **(I,A,N)**
169. Phosalone **(I,A)**
170. Phosphamidon (**(I,A)**)
171. Primiphos-methyl **(I,A)**
172. Prallethrin **(I)**
173. Pretilachlor **(H)**
174. Profenophos **(I,A)**
175. Propanil **(H)**
176. Propergite **(A)**
177. Propetamphos **(I,A)**
178. Propiconazole **(F)**
179. Propineb **(A)**
180. Propoxur **(I)**
181. Pyrachlostrobin **(F)**
182. Pyrethrins (pyrethrum) **(I,A)**
183. Pyridalyl **(I)**
184. Pyriproxyfen **(JH)**
185. Pyrithio bac sodium **(H)**
186. Quinalphos **(I,A)**
187. Quizalofop ethyl **(I)**
188. S-bioallethrin **(I)**
189. Sirmate **(H)**
190. Sodium Cyanide **(Inorganic)**
191. Spinosad **(I)**
192. Streptomycin + Tetracycline **(B)**
193. Sulfosulfuron **(H)**
198. Sulphur **(F, A)**
199. Tebuconazole **(F)**
200. Temephos **(I)**
201. Thiacloprid **(I)**
202. Thiomethoxain **(I)**
203. Thifluzamide**(F)**
204. Thiobencarb (Benthiocarb) **(H)**
205. Thiodicarb**(I)**
206. Thiometon **(I, A)**
207. Thiophanate-Methyl **(F)**
208. Triacontanol **(PGR)**
209. Triadimefon **(F)**
210. Triallate **(H)**
211. Triazophos **(I,A, N)**

212.	Trichlorofon **(I)**
213.	Trichoderma virde **(BF)**
214.	Tricyclazole **(F)**
215.	Tridemorph **(F)**
216.	Trifluralin **(H)**
217.	Validamycin **(F)**
218.	Verticillium lecanii **(Micro,I)**
219.	Zinc Phosphide **(R)**
220.	Zineb **(F)**
221.	Ziram **(F, R, Rep)**

Figures in parenthesis: A: Acaricides, Al: Algicide; B: Bactericides, BF: Biofungicide; Bio: Biopesticides F: Fungicides, Fg: Fumigant, H: Herbicides, I: Insecticides, PGR: Plant Growth Regulator, M: Mollusicide, Micro: Microbial; N: Nematicide, R: Rodenticide, Rep: Repellent; JH: Juvenile Harmone
(Source: Anonymous, 2009)

Table 1.10: Pesticides banned for manufacture, import and use (25 Nos.)

Sr. No.	Name of Pesticide	Sr. No.	Name of Pesticide
1.	Aldrin	**15.**	Pentachlorophenol
2.	Benzene Hexachloride	**16.**	Phenyl Mercury Acetate
3.	Calcium Cyanide	**17.**	Sodium Methane Arsonate
4.	Chlordane	**18.**	Tetradifon
5.	Copper Acetoarsenite	**19.**	Toxafen
6.	Clbromochloropropane	**20.**	Aldicarb
7.	Endrin	**21.**	Chlorobenzilate
8.	Ethyl Mercury Chloride	**22.**	Dieldrin
9.	Ethyl Parathion	**23.**	Maleic Hydrazide
10.	Heptachlor	**24.**	Ethylene Dibromide
11.	Menazone	**25.**	TCA (Trichloroacetic acid)
12.	Nitrofen	**26.**	Metoxuron
13.	Paraquat Dimethyl Sulphate	**27.**	Chlorofenvinphos
14.	Pentachloro Nitrobenzene		

(Source: Anonymous, 2009)

Table 1.11: Pesticide / Pesticide formulations banned for use but their manufacture is allowed for export (2 Nos.)

Sr. No.	Name of Pesticide
1.	Nicotin Sulfate
2.	Captafol 80% Powder

(Source: Anonymous, 2009)

Table 1.12: Pesticide formulations banned for import, manufacture and use (4 Nos)

Sr. No.	Name of Pesticide
1.	Methomyl 24% L
2.	Methomyl 12.5% L
3.	Phosphamidon 85% SL
4.	Carbofuron 50% SP

(Source: Anonymous, 2009)

Table 1.13: List of withdrawn pesticide (7 Nos)

Sr. No.	Name of Pesticide
1.	Dalapon
2.	Ferbam
3.	Formothion
4.	Nickel Chloride
5.	Paradichlorobenzene(PDCB)
6.	Simazine
7.	Warfarin

(Source: Anonymous, 2009)

Table 1.14: List of pesticides refused registration

Sr. No.	Name of Pesticide	Sr. No.	Name of Pesticide
1.	Calcium Arsonate	10.	Azinphos Ethyl
2.	EPM	11.	Binapacryl
3.	Azinphos Methyl	12.	Dicrotophos
4.	Lead Arsonate	13.	Thiodemeton / Disulfoton
5.	Mevinphos (Phosdrin)	14.	Fentin Acetate
6.	2,4, 5-T	15.	Fentin Hydroxide
7.	Carbophenothion	16.	Chinomethionate (Morestan)
8.	Vamidothion	17.	Ammonium Sulphamate
9.	Mephosfolan	18.	Leptophos (Phosvel)

(Source: Anonymous, 2009)

Table 1.15: List of pesticides for restricted use in India

Sr. No.	Name of Pesticide	Sr. No.	Name of Pesticide
1.	Aluminium Phosphide	8.	Monocrotophos(ban for use on vegetables)
2.	DDT	9.	Endosulfan
3.	Lindane	10.	Fenitrothion
4.	Methyl Bromide	11.	Diazinon
5.	Methyl Parathion	12.	Fenthion
6.	Sodium Cyanide	13.	Dazomet
7.	Methoxy Ethyl Mercuric Chloride (MEMC)		

(Source: Anonymous, 2009)

Part D

1.11 Present Day Pesticides

The agrochemical industry has changed significantly since the early days of pesticide development. While the older products no longer covered by patents have moved to generic companies, those investing in R&D have not only diversified into GM crops, but have also realized that the registration authorities are unlikely to accept the most toxic pesticides, nor those which are very persistent in the environment. Now, broad spectrum pesticides are needed with potential use of large areas of a major crop to cover the development costs, there has been recognition of the need for more selective products or more selective use of the broader spectrum chemicals.

1.11.1 Herbicides

Most new herbicides have been added to existing types, such as the sulfonylureas. More refined production has led to the δ-isomer of metolachlor replacing the earlier product. In general, new herbicide products are often different combinations of herbicides in pre-mixed formulations to suit specific weed situations in different crops and countries. There has also been greater awareness of enhancing herbicide activity by recommending the addition of certain adjuvants, such as methylated vegetable seed oils, that improve the spread of the spray deposit on foliage and increase the amount taken up by the weeds.

1.11.2 Fungicides

The strobilurins, synthetic analogues of strobilurin A produced by *Strobilurus tenacellus,* have been the main new group of fungicides, although new versions of older groups continue to be developed. Prothioconazole is a new azole (Mauler-Machnik et al., 2002). Ethaboxam is a new fungicide which is specific to controlling oomycetes such as grape downy mildew, and late blight on potato (Kim et al., 2002).

1.11.3 Insecticides

In place of many of the organochlorine, organophosphate and carbamate neurotoxic insecticides, new groups include the pyrethroids and nicotinoids. Other new groups are chitin synthesis inhibitors and other insect growth regulators as well as avermectin, milbemectin and certain other new pesticides, such as pyrazoles.

1.11.4 Neonicotinoids

The more recent development is this group of insecticides, which emulates the effect of nicotine derived from tobacco. These insecticides, such as imidacloprid, act on the acetylcholine system by blocking the postsynaptic nicotinergic acetylcholine receptors. The use of imidacloprid has expanded rapidly since 1990, both as a spray and seed treatment to utilise systemic activity against a range of pests including aphids and whiteflies. Similarly, thiamethoxam is recommended for control of these pests. Another new neonicotinoid is chlothianidin (Ohkawara et al., 2002). The risk is that overuse of this group of insecticides will lead to widespread resistance.

1.11.5 Phenylpyrazole

Fipronil is an example of another new type of broad-spectrum insecticide used at very low dosages. It is also persistent and must be used with caution. It is nevertheless an important tool in some pest control situations as it provides a different mode of action.

1.11.6 Insect Growth Regulators (IGRs)

In contrast to the neurotoxic poisons, these chemicals interfere with the growth of the immature stages of insects. At the larval stage of an insect, it moults and forms a new skin or cuticle. This process is controlled by hormones and requires the production of chitin that forms the skin of the next larval stage. IGRs can be categorised into three main groups: juvenile hormone analogues; anti-juvenile hormones; and chitin synthesis inhibitors. The latter group has been the most widely used and includes diflubenzuron. Larvae affected by this insecticide will start to moult into the next instar but fail to complete the process. The action is slow as no effect is discernible until the insect moults. Adults are not killed, but there is some evidence that oviposition is adversely affected if they contact a sufficient dose. These insecticides have an extremely low toxicity to mammals.

Juvenile hormone analogues, such as methoprene have been used to control mosquito larvae and have also been successful in controlling Pharaoh's ants in buildings, although the effect is not seen for some time after the application. Tebufenozide is an example of the anti-juvenile hormone type, which cause larvae to form precocious adults.

1.11.7 Spinosads

In 1982, a new species of actinomycete was found in a soil sample from the Caribbean. From this *Saccharopolyspora spinosa* two fermentation products led to the development of new class of insecticide, the naturalyte class. Spinosad is the first product to be commercialised in this class. Its mammalian toxicity and environmental profile make it an excellent insecticide in IPM programmes. Spinosad is degraded by sunlight, but surface deposits become stabilised with activity at a range of pH values so it remains sufficiently effective on foliage to control a range of lepidopteron pests, yet it is safe to most beneficials. It has already been used extensively in conventional cotton crops to replace pyrethroids and to supplement control on *Bt* cotton.

1.11.8 Other Insecticides

The quest for new compounds with different modes of action continues. Pyridalyl is a new insecticide with low mammalian toxicity, yet good activity against lepidopteran pests (Saito et al., 2002). Another group, the spirocyclic phenyl-substituted tetronic acids, is showing promise with spiromesifen having activity against whiteflies and spider mites (Nauen et al., 2002*)*. Spirodiclofen has activity against mites and psyllids and scale insects, and offers an important tool in IPM fruit production (De Maeyer et al., 2002*)*. Another new systemic insecticide of low mammalian toxicity is flonicamid, reported to have a different mode of action and be effective as an alternative to organophosphates and neonicotinoids.

Conclusion

Pesticides are indispensable for feeding, clothing and protecting the world population from diseases and discomfort. Pesticides are one spoke in the wheel that makes up IPM (Integrated Pest Management). Search for new plant protection techniques will continue for ever for sustainable agricultural production and safety of consumers and environment. Recently, some new pesticides have been developed which are biodegradable, high mammalian safety, low residual life and compatible with non-target organisms. The trend will continue for evolving products having much more specific and less broad spectrum in nature targeting specific pests or pest families and using greatly at low rates of application.

References

Anonymous, 1954. *Chronological History of the Development of Insecticides and Control Equipment from 1854 through 1954.* USDA Agricultural Research Service, USA.

Anonymous, 1999. *Organophosphates: Committee on Toxicity of Chemicals in Food, Consumer Products and the Environment.* Department of Health, London, UK.

Anonymous, 2002. *37th Proceeding of Ministry of Chemicals & Petrochemicals, Govt. of India.*

Anonymous, 2005. *PDP Pesticide History, USDA, AMS, S & T, MPO – Pesticide Data Program (PDP). The Pesticide Book, 6th edn.* (2004). Meister Pro Information Resources, Ohio, USA.

Anonymous, 2005a. *States/UTs (Zonal Conferences on Inputs, Plant Protection (PP), Kharif.)*

Anonymous, 2008. Average crop losses due to pest & cost benefit ratio of pesticides use in different crops. In: *Crop Care, Official Magazine of Crop Care Federation of India* (ed. Kishore, P.): 75pp.

Anonymous, 2008a. *Central Insecticide Board & Registration Committee.* Ministry of Agriculture, India.

Anonymous, 2009. *Directorate of Plant Protection, Quarantine and Storage.* Faridabad, Haryana, India, 2009.

Carson, R., 1962. *Silent Spring.* Hamish Hamilton, UK.

Casida, J.E., 1973. *Pyrethrum, The Natural Insecticide.* Academic Press Inc. London., UK.

De Maeyer, I., Peeters, D., Wijsmuller, J. M., Cantoni, A., Brueck, E., and Heibges, S., 2002. Spirodiclofen: A broad-spectrum acaricide with insecticidal properties: Efficacy on Psylla pyri and scales Lepidosaphes ulmi and Quadraspidiotus perniciosus. Proc. *Brit. Crop Protec. Conf. on Pests and Diseases,* Brighton, UK: 65-72.

Dunne, B., 2002. *New Fungicides and Their Role in Disease Control Programmes.* Teagasc, Crops Research Cetnre, Oak Park , Carlow, Ireland.

Eto, M., 1974. *Organophosphours Pesticides: Organic and Biological Chemistry.* CRC Press Inc. Ohio, USA.

FAO, 1986. *International Code of Conduct on the Distribution and Use of Pesticides.* Food and Agriculture Organization of the United Nations, Rome, Italy: 31 pp.

Fest, C. and Schmidt, K.J., 1973. *The Chemistry of Organophosphorus Pesticides.* Springer-Verlag, Berlin, New York, USA.

Handa, S.K., 1999. *Principles of Pesticide Chemistry.* Agrobios Publishers, New Delhi, India.

Hewitt, H.G., 1998. *Fungicides in Crop Protection.* CAB International, Wallingford, Oxon, UK, New York, USA.

Karalliedde, L., Feldman, S., Henry, J., and Marrs, T., 2001. *Organophosphates and Health.* IC Press, London, UK.

Kim, D.S., Lee, Y.S., Chun, S.J., Choi, W.B., Lee, S.W., Kim, G.T., Kang, K.G., Joe, G.H., and Cho, J.H., 2002. Ethaboxam: A new oomycetes fungicide. Proc. *Brit. Crop Protec. Conf. on Pests and Diseases,* Brighton, UK: 377-382.

Kuhr, R. J. 1971. The formation and importance of carbamate insecticide metabolites as terminal residues. In: *Pesticide terminal residues* (ed. Tahori, A. S.), Butterworths, London: 199-220.

Kuhr, R.J., and Dorough, H.W., 1976. *Carbamate Insecticides: Chemistry, Biochemistry, and Toxicology,* CRC Press, Inc. Ohio, USA.

Lock, E.A. and Wilks, M.F., 2001. Paraquat: In: *Handbook of Pesticides Toxicology*, 2nd edn.(ed. Kriger, R.I.), Academic Press, San Diego, USA:1559-1603.

Matsmura, F., 1985. *Toxicology of Insecticides* 2nd edn., Plenum Press, New York, USA.

Mauler-Machnik, A., Rosslenbroich, H. J., Dutzmann, S., Applegate, J., and Jautelat, M., 2002. JAV 6476 - A New Dimension DMI Fungicide. Proc. *Brit. Crop Protec. Conf. on Pests and Diseases,* Brighton, UK: 389-394.

Nauen, R., Bretschneider, T., Bruck, E., Elbert, A., Reckmann, V., Wachendorff, V., and Tiemann, R., 2002. BSN 2060: A novel compound for whitefly and spider mite control. *Proc. Brit. Crop Protec. Conf. on Pests and Diseases,* Brighton, UK: 39-44.

Ohkawara, Y., Akayama, A., Maysuda, K., and Andersch, W., 2002. Clothianidin: A novel broad-spectrum neonicotinoid insecticide. Proc. *Brit. Crop Protec. Conf. Proc. Brit. Crop Protec. Conf. on Pests and Diseases,* Brighton, UK: 51-58.

Pimental, D., 1987. Energy use in chemical agriculture. In: *Towards a second green revolution* (eds. Marini-Bettelo, G.B.), Elsevier, Amsterdam, USA: 157-176.

Saito, S., Isayama, S., Sakamoto, N., Umeda, K. and Kasamatsu, K., 2002. Pyridalyl: A novel insecticidal agent for controlling lepidopterous pests. Proc. *Brit. Crop Protec. Conf. on Pests and Diseases,* Brighton, UK: 33-38.

Schrader, G., 1942. *German Patent,* 720577.

Schrader, G., 1963. *Die Entwicklung uber insektizider Phosphoraure-Ester.* Verlag Chemie, Weinheim, Germany.

Singh, B., and Dhaliwal, G.S., 2000. Pesticide contamination of fatty food commodities. In: *Pesticides and environment* (eds. Dhaliwal G.S. and Singh, B.), Commonwealth Publishers, New Delhi: 155-190.

Tomizawa, M., and Casida, J.E., 2005. Neonicotinoid insecticide toxicology: Mechanism of selective action. *Ann. Rev. of Pharmaco. Toxicol.* **45:** 247-68.

Ware, G.W., and Whitcare, D.M., 2004. *The Pesticide Book,* 6th edn., Messterpro Information Resources, Willoughby, Ohio, USA.

❑❑❑

Chapter 2

Pesticide Metabolism in Insects, Plants and Animals

The knowledge on the fate of pesticides, after application in the environment is useful to research workers associated with their safe and effective use. It is important to know whether a xenobiotic will persist or will be inactivated. Frequently persistence is desirable for sustained control. However, for many cases, rapid detoxification of the compound may be required to avoid cumulative effects. Most of the pesticides are not very soluble in water and their oxidation or hydrolysis (or both) leads to formation of polar groups. Such reactions are called primary metabolic reactions, often, but not always. The products of these primary reactions undergo secondary changes whereby they are conjugated with endogenous substances to form molecules that are more readily excreted. In higher animals, these changes facilitate the transportation of metabolites of pesticides in blood, which in turn leads to their eventual excretion.

Normally, lipophilic xenobiotics that enter an animal's body are rapidly detoxified. Biotransformation processes of pesticides are grouped into three main phases; known as phase - I or conversion, phase-II or conjugation and phase-III or compartmentation (Figure 2.1). Phase-I reactions include oxidations, reductions and hydrolysis, whereas phase-II reactions are conjugations with glutathione, sugars or amino acids. In phase-III, xenobiotic conjugates are converted to secondary conjugates or insoluble bound residues and are deposited in the vacuole or other compartment of the plant cells. The biotransformation of xenobiotics is dependent upon the organism species, giving rise to pesticide detoxification and activation, or herbicide selectivity and resistance. The most important function of biotransformation is to decrease the lipophilicity of xenobiotics, so that ultimately they can be excreted. The following references will be useful for understanding detoxification mechanisms in animals, especially insects: Kuhr,(1970), Nakatsugawa and Morelli (1976), Yang (1976), Dauterman (1985), Hodgson (1985), Ahmad et al. (1986), Hodgson and Levi (2001), Matsumura (1985), Yu (2004), Li et al., (2007) and Yu, (2008).

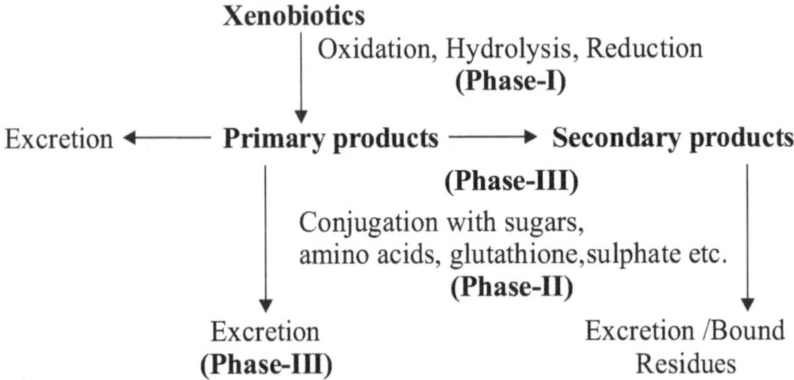

Figure 2.1: Schematic representation of xenobiotics metabolism

2.1 Significance of Metabolism

Before discussing details about the metabolic processes of pesticides and the enzymes involved, it is necessary to consider as to why this subject is of fundamental importance to all concerned with chemicals. In the first instance, metabolism plays an important role in relation to selectivity of action of pesticides and thus plays a part in determining the safety of compounds to man and farm animals. Secondly, the level of metabolism is a factor determining the persistence of pesticides in soil, plants and animal bodies, and therefore contributes significantly to the reduction of environmental pollution. Finally, metabolic factors are frequently implicated in the mechanism leading to the development of pest resistance. Additionally, knowledge of metabolism is essential for developing analytical techniques for estimation of residues which can detect and estimate the parent moieties and their toxic metabolites.

2.1.1 Different Metabolic Pathways of Insecticides

Metabolic pathways can be classified into seven groups that cover the majority of biotransformation which pesticides undergo. These are:

1. Oxidation (in the sense that oxygen, as an hydroxyl takes part or is postulated to take part in one or more of the steps).
   (a) Hydroxylation of aromatic rings
   (b) Oxidation of the side chains to alcohols, ketones or carboxyl groups
   (c) Dealkylation from oxygen or sulfur (ether cleavage)
   (d) Sulfoxide formation
   (e) N-oxide formation

2. Dehydrogenation and dehydrohalogenation
3. Reduction
4. Conjugation
   (a) Amide formation
   (b) Metal complex
   (c) Glucoside or glucuronic acid
   (d) Sulfate
6. Exchange reactions
7. Isomerisation

2.1.2 Enzymes Involved in Metabolism of Insecticides

2.1.2.1 Hydrolases (Esterases)

In very broad sense these are enzymes which hydrolyze carboxylic and phosphorus esters. The esterases abbreviated as it include the aliesterase, and a great variety of other enzymes hydrolyzing phosphours esters. It also includes the diesterase. Esterases can be broadly derived in to two groups termed as A esterases and B esterases. A esterases are able to hydrolyze organophosphates such as paraoxon at ordinary concentration levels, whereas B esterases are under the same conditions inhibited by organophosphates. A metal ion frequently is an activator for the esterases which in mammals are very active in plasma, liver and kidney.

2.1.2.2 Microsomal Polysubstrate Oxygenases (Mixed Function Oxidase i.e. MFO system)

These enzymes insert one of the two oxygen atoms from an oxygen molecule into an appropriate substrate, RH. The other oxygen atom eventually forms part of a water molecule.

$$RH + O_2 + [2H] \rightarrow R\text{-}OH + H_2O$$

A unique feature of the major mono oxygenase system is presence of a haem protein termed cytochrome P_{450} which links electron flow to the reduction of the oxygen molecule. The cytochrome P_{450} system has an absolute requirement for NADPH. In mammals, these enzymes reside largely in the endoplasmic reticulum or microsomal fraction of liver while in insects they are concentrated in the fat body, intestine and malpighian tubules. The main features of the mono oxygenase system are summarized in Figure 2.2.

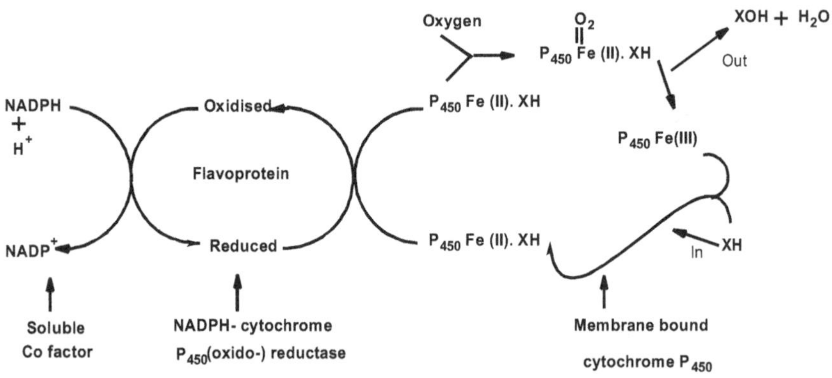

Figure 2.2: Proposed schemes for the oxidation of a foreign compound, XH, by the cytochrome P $_{450}$ mono-oxygenase system.

(Source: Hassall, 1990)

The substrate, XH, probably gets attached to the ferric form of cytochrome P_{450}. An electron from NADPH reduces the complex to the ferrous form. The oxidising agent, molecular oxygen, then attaches itself to the ferrous form of a cytochrome P_{450} substrate complex. The bond between two oxygen atoms is then weakened, possibly by addition of a second electron. An intermolecular electron shift leads to the formation of peroxide - XH - ferric P_{450} complex. This apparently loses an oxide ion to protons in the medium, leaving behind a highly active oxygen atom, which reacts with the XH to form, X-OH. The ferric form of cytochrome P_{450} is then ready to accept more substrate. The enzyme that passes electron from NADPH to cytochrome P_{450} is a flavoprotein. Thus, the microsomal electron transport system comprises at least three components, namely a reduced pyridine nucleotide a flavoprotein and cytochrome.

2.1.2.3 Glutathione-S- Transperase

Glutathione (GSH) is a tripeptide comprising residues of glycine, cysteine and glutamic acid. It is important in relation to the degradation of pesticides for a major metabolic pathway involves glutathione-S- transperase reactions. Glutathione often forms a conjugate with the intruding pesticide, whereas most other conjugation of endogenous substances take part in secondary reactions, which only occur after the original substance has undergone oxidation or hydrolysis. Glutathione conjugates can themselves undergo secondary changes, for excreted products in many organisms are not glutathione conjugates as such, but derivatives called mercapturic acids. Several groups of glutathione-S-transferases are known of which the three shown below are of considerable importance in pesticide detoxification.

(i) Glutathione-S-epoxide transferase

(ii) Glutathione-S-aryl transferase

(iii) Glutathione-S-alkyl transferase

These transferases are present in the soluble cell fraction of mammalian liver. Clark et al., (1984) achieved up to 100 fold purification of glutathione-S-transferases from houseflies. Glutathione being a tripeptide conjugates with foreign compounds to give complex of the sort shown in Figure 2.3. If the glycine and glutamic acid are hydrolytically removed, a cysteine derivative of the foreign compound remains. Acetylation of the amino group of the cysteine then frequently occurs so that the excreted material in many higher animals is an N-acetylsteine derivative. This is called mercapturic acid (derivative of glutathione). In insects, glutathione conjugates tend to be either excreted unchanged or converted to the cysteine rather than to acetylcysteine derivatives (Walker, 1975).

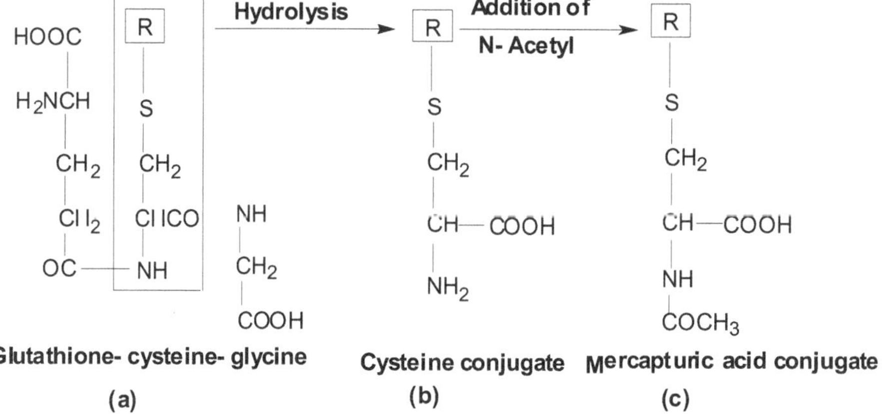

Figure 2.3: Structure of glutathione and mercapturic acid conjugates
(a) Glutathione, a tripeptide conjugates with a pesticide, R; (b) A cysteine is formed on hydrolysis of (a); (c) N-acetylation of (b) gives a mercapturic acid derivative.

(Source: Hassall, 1990)

2.2 Metabolism of Organochlorine Insecticides

The insecticidal potential of certain organochlorine insecticides (OC) was discovered during the Second World War. DDT and BHC were first praised as almost perfect insecticides, being relatively cheap, highly effective against a variety of insect-pests and yet relatively safe to man and other warm blooded animals. Out of a large number of OC compounds, two have been taken for knowing their metabolic behaviour in various matrices.

2.2.1 DDT

There are five principal routes of DDT metabolism (Figure 2.4) in various organisms:

(i) Oxidation to DDA
(ii) Oxidation to kelthane
(iii) Oxidation to dichlorobenzophenone
(iv) Dehydrochlorination to DDE
(v) Reductive dechlorination to DDD

DDE is the major DDT-derived residue normally found in autopsy or biopsy samples of animals tissues (Robinson, 1969). The enzyme involved in this conversion has been termed DDT dehydrochlorinase or DDTase. The reaction is dependent on presence of glutathione and, at least in houseflies, it is brought about by a glutathione -S - transferase (Clark and Shamaan, 1984).The development of resistance to DDT in some species of houseflies and mosquitoes is closely correlated with the greater capacity of resistant organisms to convert DDT to the less insecticidal DDE. A few insects metabolize DDT by a predominantly oxidative route, which results in the formation of dicofol (Kelthane). The route seems to be a typical NADPH - dependent microsomal oxidation. This aliphatic hydroxylation occurs in *Drosophila* and perhaps in resistant strains of the cockroach. Reductive dechlorination of DDT occurs widely in nature and DDD is an almost ubiquitous residue in animal fat (Robinson, 1969).

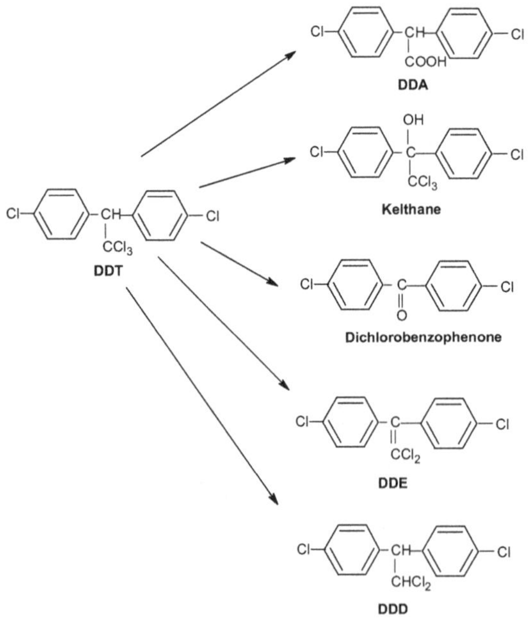

Figure 2.4: Metabolic pathways of DDT

In mammals, a water soluble metabolite, DDA is formed by the replacement of the – CCl_3 part of DDT molecule by -COOH. This acid may be excreted in free or conjugated form in faeces or in bile according to the species. It is not clear by what route or routes DDA is formed, in particular it is not clear whether DDD is an obligatory intermediate or whether it can be formed via DDE.

2.2.2 Endosulfan

Endosulfan is a cyclic sulphite and is thus not a true cyclodiene, although the characteristic chlorinated bridged ring is present. After deposition on a plant, endosulfan changes to its oxidation product, endosulfan sulfate which is very toxic to higher organisms. It is the principal metabolite produced by some fungi, although other hydrolyse endosulfan to give a diole. The other metabolites reported are endosulfan hydroxy ether, and endosulfan lactone (Figure 2.5).

Endosulfan → (Oxidation, mfo) → Endosulfan sulfate

Endosulfan → (Hydrolysis) → Endosulfan diol → (-HOH) → Endosulfan ether → (Side chain Hydroxylation) → Hydroxy endosulfan ether ⇄ (Oxidation) Endosulfan lactone

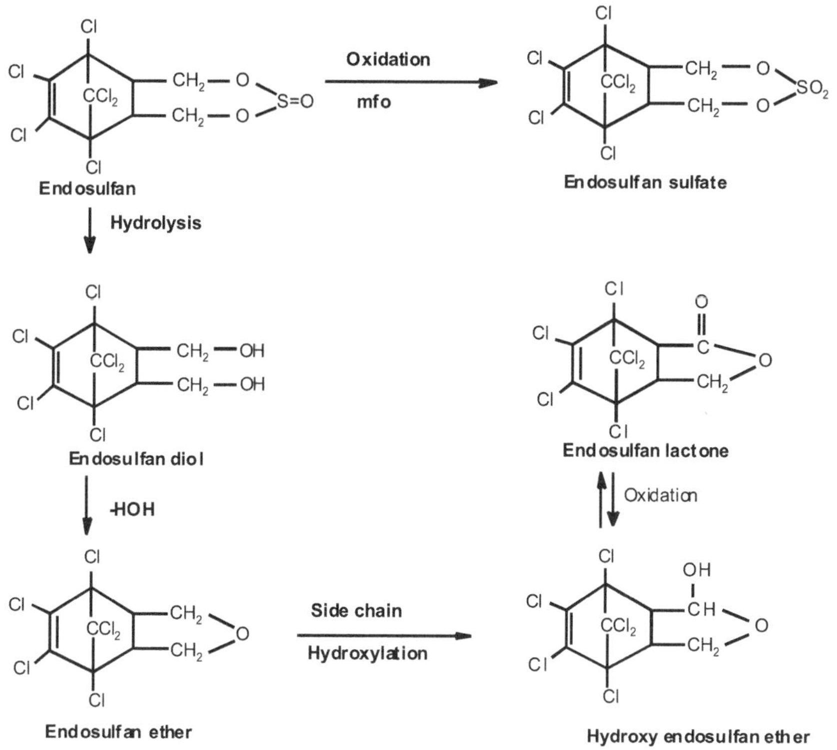

Figure 2.5: Metabolic pathways of endosulfan

The toxicity of endosulfan sulfate to mammals is about the same as for the parent compound itself, whereas the diol, the hydroxy ether and lactone ranges from 150-15000 mg/kg for rat.

2.3 Metabolism of Some Organophosphate Compounds

2.3.1 Parathion, Methyl Parathion and Fenitrothion

The three principal members of this family are parathion, methyl parathion and fenitrothion. The metabolism of parathion and its two analogues methyl parathion and fenitrothion illustrates (Figures 2.6 and 2.7) how closely field safety is associated with minor variations in chemical structure, which, in turn, lead to changes in the balance established between the various pathways of metabolism. All three compounds are oxidised by mono-oxygenases in animals, insects and plants and are thereby changed to derivatives containing the P=0 group, which are more powerful inhibitors of cholinesterase than the original thionophosphate. Their degradation, however, follows different routes. Parathion is principally degraded by rupture of the P-O-phenyl linkage, whereas the methyl analogues are not. The rupture of the P-O-phenyl linkage of parathion is not a hydrolytic reaction but is an oxidative process mediated by an NADPH-dependent oxidase. By contrast, methyl parathion and fenitrothion are principally destroyed in most species investigated by rupture of P-O-CH_3, linkage.

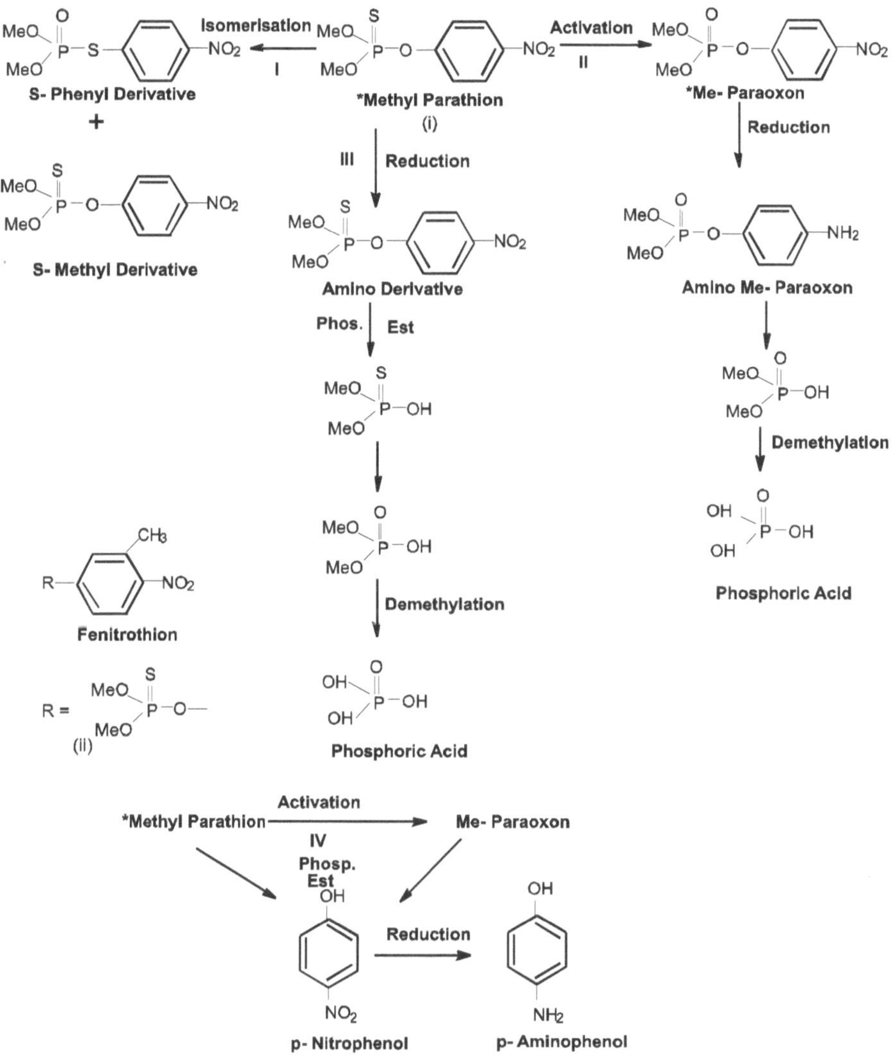

Figure 2.6: Metabolic pathways of thiophosphates, (i) Me-parathion (ii) Fenitrothion

(Source: Kathpal and Yadav, 2000)

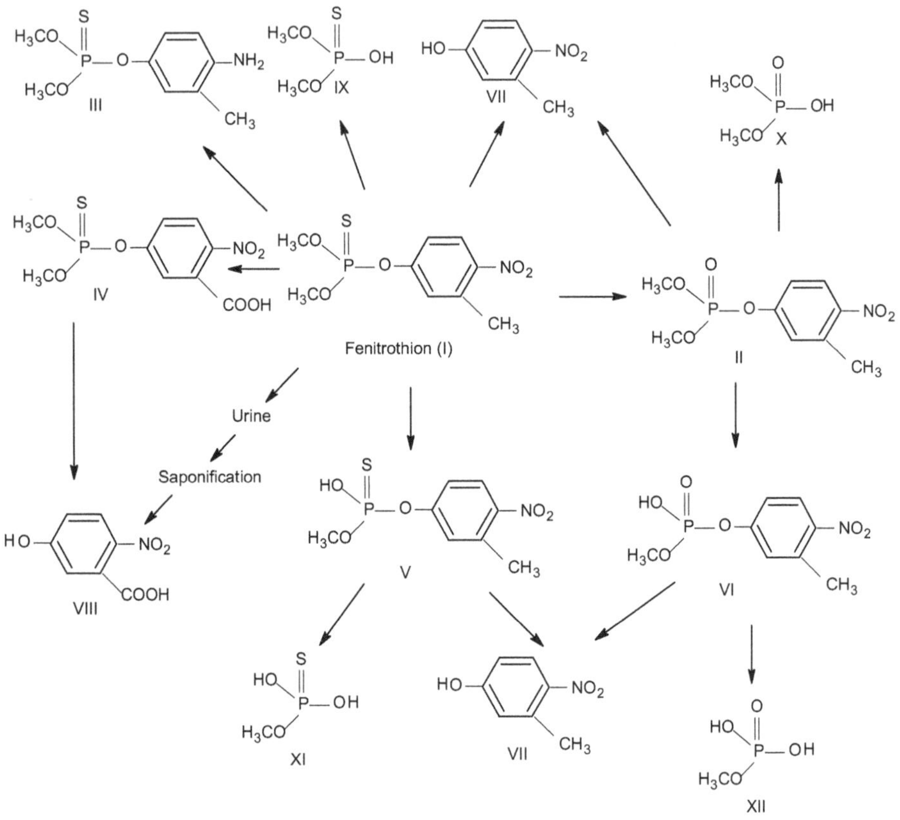

Figure 2.7: Metabolism of fenitrothion

(Source: www.inchem.org)

2.3.2 Malathion

It is a dithiophosphate having succinic acid ester as leaving group. Malathion provides an elegant example of the way selective action can sometime be achieved by exploiting differences in the enzymic constitution of different organisms. Almost always, hydrolysis leads to detoxification, so organisms in which a carboxyesterase is particularly active are able to protect themselves from malathion by removing one of the ethyl groups from the succinic ester moiety. Most vertebrates seem to be well endowed with this enzyme. However, when the thion sulphur ($P=S \rightarrow P=O$) is replaced by oxygen, a much more powerful cholinesterase inhibitor, malaoxon, is formed. It so happens that most insects seem to possess a very active oxidative enzyme system. This can activate malathion by attacking the P-S linkage, so that the insect's own defence system actually helps to kill it. It is stressed that all insects and all vertebrates probably possess both an esterase and an NADPH-dependent oxidase system, it is the balance of the action of these two systems that varies from one organism to another. Further metabolic changes can remove both ethyl groups from the succinate moiety. There is also some evidence that a glutathione-S-methyl transferase can remove methyl groups from one or both of the two methoxy groups (Figure 2.8)

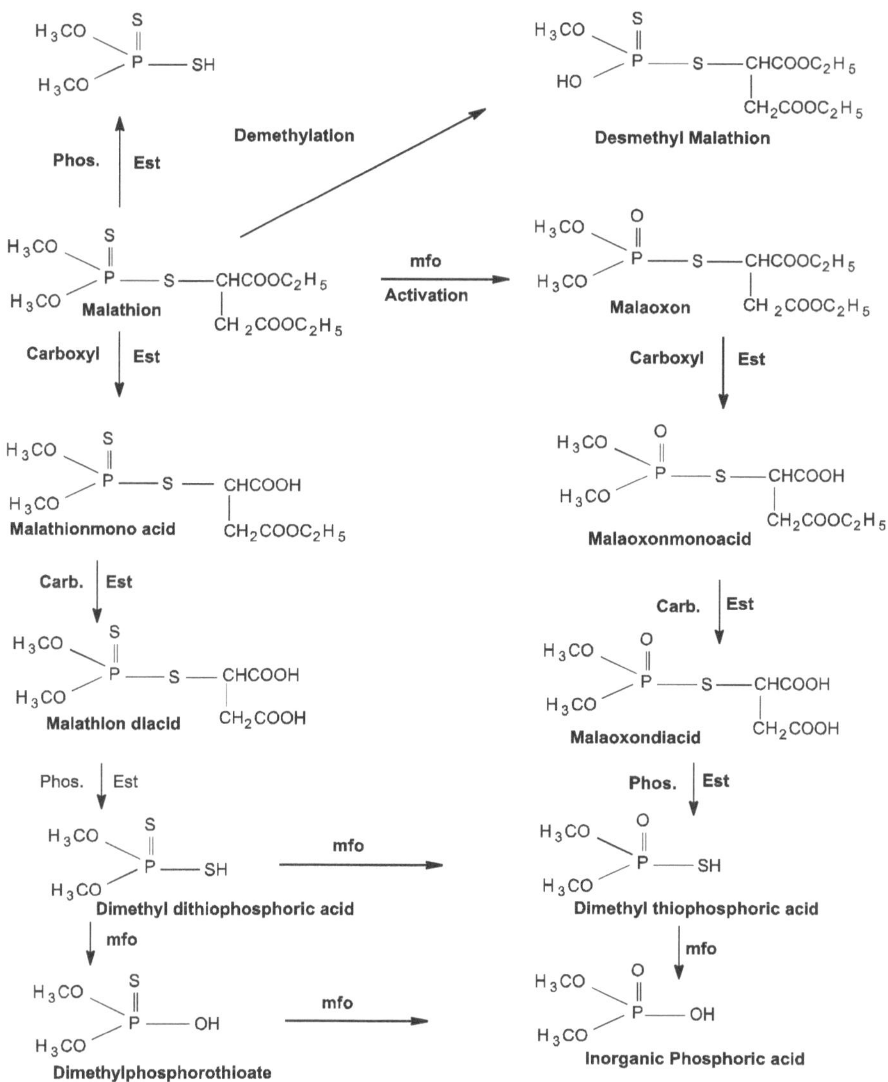

Figure 2.8: Metabolic pathways of malathion

(Source: Kathpal and Yadav, 2000)

2.4 Metabolism in Carbamate Insecticides

Carbamate insecticides are frequently employed to control insects that, for some reason do not readily respond to organophosphorous compounds. They usually control more susceptible insects as well. Their higher cost is usually a reason to regard them as essentially 'heavy duty' insecticides. As carbamates find their use on a variety of crops it becomes imperative to have clear understanding regarding their metabolism in various ecosystems. Metabolic pathway of two representative carbamate insecticides has been discussed.

2.4.1 Carbaryl

This insecticide is used in greatest amount all over the world against at least 150 major pests of crops. Hydrolysis appears to play a less important initial role in metabolism of carbamates than it does for organophosphates. The extent of hydrolysis varies from one species to another, a considerable proportion of a dose of carbaryl is hydrolyzed by the rats and German cockroach but not by pig or by American cockroach (Kuhr, 1971). Plants do not appear to hydrolyse it significantly. Some species of bacteria tend to metabolise it by oxidation while others appear to do so largely by hydrolysis. In carbaryl, (Figure 2.9) oxidative routes lead to 4-hydroxy and 5-hydroxy derivative of carbaryl. In addition, epoxidation and N-hydroxy methyl carbaryl formation are characteristic of reactions normally catalysed by NADPH-dependent mono-oxygenases. Hydrolysis of carbaryl epoxide by an epoxide hydrolase leads to the formation of a diol. Since hydrolysis of the carbaryl ester linkage may occur more rapidly after initial oxidation in ring positions 4 and 5, the isolation of 1, 4-or 1, 5-dihydroxynapthols does not necessarily imply that the initial attack on the molecule is hydrolytic. Hydroxy compound whether formed by oxidation or by hydrolysis can be conjugated in a secondary stage, which always results in detoxification. Pekas (1979) demonstrated that the small intestine of the rat converted carbaryl to naphthyl glucuronide and to hydroxy naphthyl glucuronide.

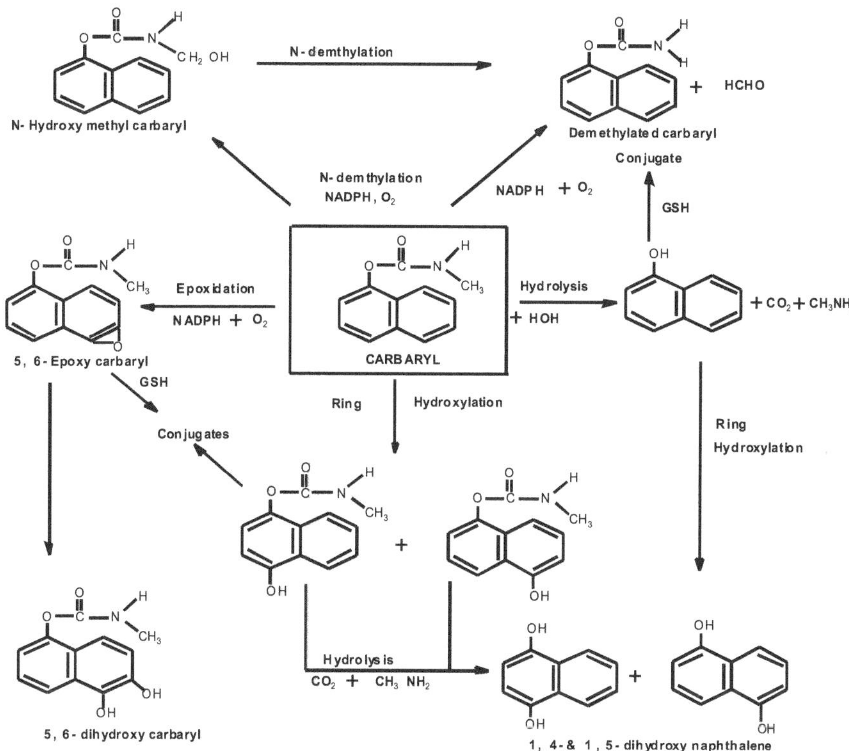

Figure 2.9: Metabolic pathways of carbaryl
(Source: Hassall, 1990)

2.4.2 Aldicarb (An Oxime Carbamate)

Aldicarb or Temik is an oxime insecticide. It is one of the most potentially toxic substances currently used in crop protection with an oral LD_{50} (Rats) of about 1 mg/kg. Its metabolic fate is different than that of the carbamate insecticides because it possesses an extra hydrolyzable bond. Since it is also a thioether, oxidation of aldicarb to a sulfoxide analogue is expected. Hydrolysis, which follows the oxidative reaction, probably plays the second most important role in mammalian systems (Knaak et al., 1966; Andrawes et al., 1967). Metabolism in aerobic surroundings appears to be similar in animals and in soil (Figure 2.10). In animal liver, preparation of the principal primary metabolite is the sulfoxide. A small proportion of the latter is normally further oxidised to the sulfone, although conversion of the amide moiety of the sulfoxide to a sulfoxide nitrile has also been reported (Metcalf et al., 1966; Coppedge et al., 1967; Bartley et al., 1970). In consequence, the bulk of the identified metabolites often comprise the sulfoxide, its hydrolysis products and the hydrolysis products of the sulfoxide nitrile (Kuhr and Dorough, 1976). In anaerobic soil containing ferrous, iron, aldicarb is normally degraded within a few hours to give a nitrile and aldehyde (Bromilow et al., 1986).

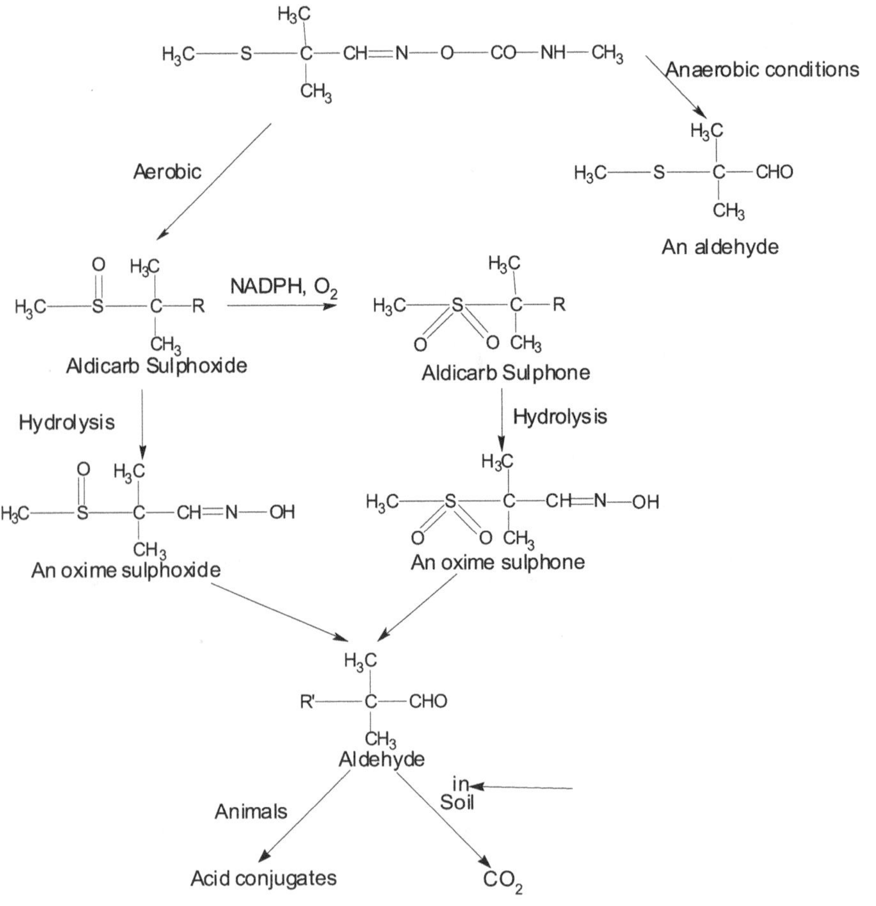

Figure 2.10: Metabolic pathway of aldicarb
(Source: Hassall, 1990)

2.4.3 Carbofuran

Carbofuran has LD_{50} of about 11mg/kg body weight and is 50 times more poisonous to vertebrates than carbaryl. When it is applied in soil, microbial participation in the degradation of pesticides generally involves the utilization of various enzyme systems to transfer the pesticide in to carbon and energy source for growth. It is metabolized by soil microorganisms via hydrolysis to phenolic metabolites, and oxidation to hydroxycarbofuran and then ketocarbofuran. It is not known that these phenolic metabolites (Figure 2.11) are degraded to CO_2 and H_2O (Ou et al., 1982). In animals and plants, these phenolic and hydroxylated metabolites are further converted to form glycosides, glucuronides or sulfates (Menzie,1974).

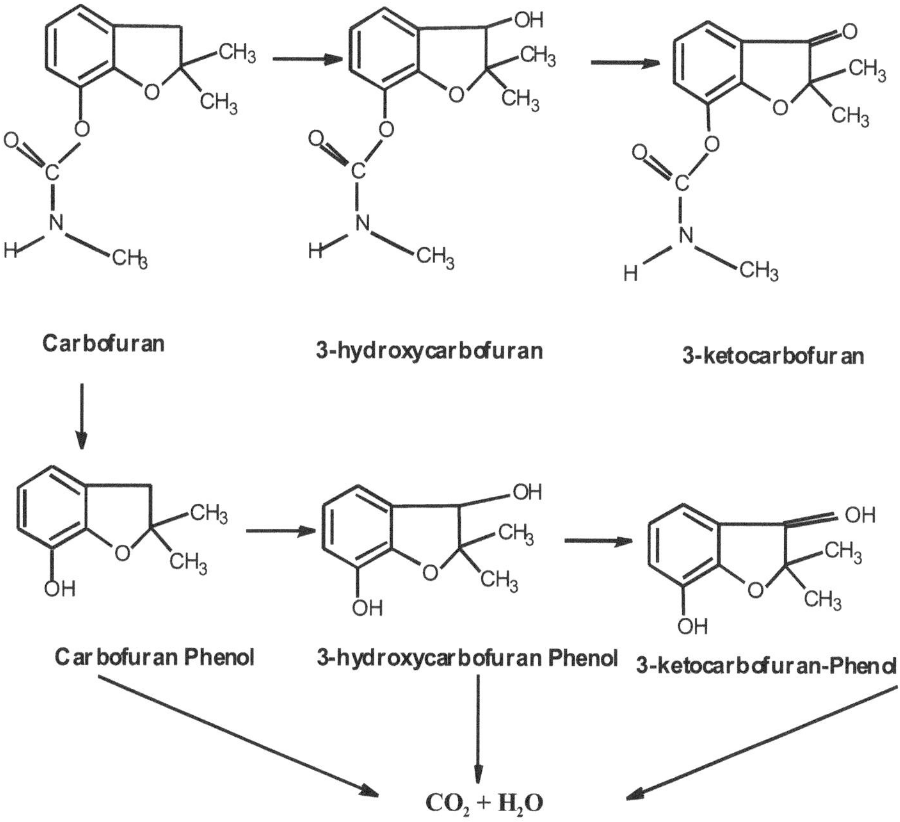

Carbofuran → 3-hydroxycarbofuran → 3-ketocarbofuran

Carbofuran Phenol → 3-hydroxycarbofuran Phenol → 3-ketocarbofuran-Phenol

$CO_2 + H_2O$

Figure 2.11: Metabolic pathway of carbofuran

(Source: Ou et al., 1982)

2.5 Metabolism of Synthetic Pyrethroids

The elucidation of structure of the natural pyrethroids made possible the synthesis of related compounds which, while possessing similar or higher insecticidal activity than natural compounds, are also more stable to light and to air. Some of the numerous compounds that have been tested and widely used, in India, are permethrin, cypermethrin, deltamethrin, fenvalerate and fluvalinate. A few of synthetic pyrethroid compounds are much more toxic to the housefly and many other insects than the natural pyrethrins. The synthetic pyrethroids are much cheaper and more specific in their action than the natural pyrethrin compounds. Metabolic pathway of only cypermethrin which is one of the highly effective and widely used chemicals of synthetic pyrethroid group has been explained below.

2.5.1 Cypermethrin

This substance is structurally similar to permethrin except that it possesses an α-cyano group at methyl group of benzyl alcohol. This insertion improves photostability and leads to a powerful and rapid debilitating effect on insects, it also greatly increases the number of possible metabolites. Cypermethrin exists as mixture of cis and trans-isomers. Croucher et al., (1985) used ^{14}C-labelled cypermethrin to study its secretion in lactating cows. After applying 0.2 to 10 ppm in food, less than one per cent was eliminated in milk, mostly in the form of unchanged cypermethrin. References to similar studies in cows and other animals are quoted by Hutson and Stoydin (1987). When ^{14}C-labelled cypermethrin was fed to domestic hens, most of the retained radioactivity was found in liver, either in the form of cypermethrin or as highly polar metabolites. Radioactivity in eggs was largely confined to the yolk (Hutson and Stoydin, 1987). The degradative metabolism of cypermethrin appears in most species to be initiated by cleavage of ester linkage of cyclopropyl carboxylic acid and phenoxy cynano benzyl alcohol or by hydroxylation in positions 2 or 4 of the ring-A of phenoxy cyano-benzyl alcohol moiety. Often, on cleavage, the alcohol loses at some stage the elements of HCN, to give 3- phenonybenzaldehyde and then 3- phenoxy acid. Tolerance to cypermethrin shown by some strains of whitefly (*Bemisia tabaci*) could probably be attributed to the presence of large amounts of an esterase that is able to destroy the pyrethroid. The major degradation pathway of cypermethrin (Figure 2.12) is hydrolysis of the ester linkage to give 3-phenoxybenzoic acid and 3-(2, 2-dichlorovinyl)-2, 2-dimethylcyclopropanecarboxylic acid. A minor degradative route is ring hydroxylation to give an α-cyano-3 (4-hydroxyphenyl) benzyl ester followed by hydrolysis to produce the corresponding hydroxycarboxylic acid.

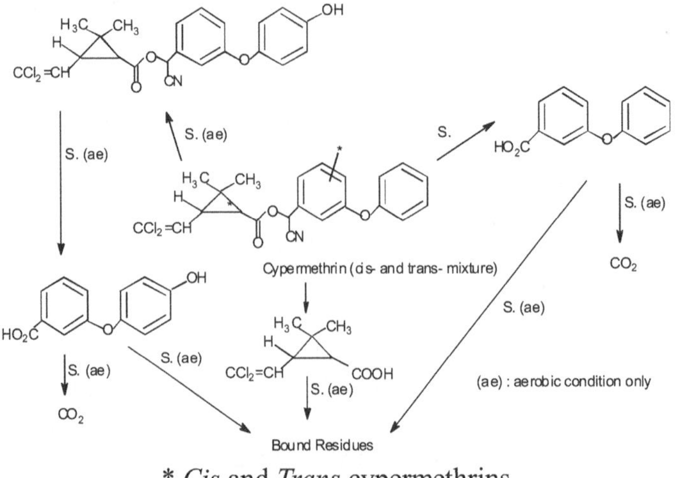

* *Cis* and *Trans* cypermethrins

Figure 2.12: Metabolic pathway of cypermethrin

(Source: Aizawa, 1982)

2.5.2 Deltamethrin

Originally called decamethrin, this substance resembles cypermethrin, it possesses a cyclopropane ring and an α-cyano group in its molecule, but differs in that it has two bromine atoms in the vinyl side chain. It is of medium to low toxicity to most mammals and birds yet kills insects at very low dosage both by contact action and by ingestion. The initial step in metabolism in most organisms appears to be by hydrolysis, leading to the formation of dibromovinyl dimethyl cyclopropane carboxylic acid and phenoxy benzyl cyanohydrins. The cyanohydrin is unstable and breaks down probably by hydrolysis to carboxyl, after loss of carbon dioxide, eventually forms 3-phenoxy benzoic acid. All of these compounds together with hydroxylated derivatives formed by the mono-oxygenase system can undergo conjugation varying with species. Deltamethrin is a cis-isomer and an extra feature of its metabolism is that it undergoes slow conversion after spraying to less toxic trans-isomer. Metabolic pathway of deltamethrin in soil and plants is shown in Figures 2.13 and 2.14, respectively.

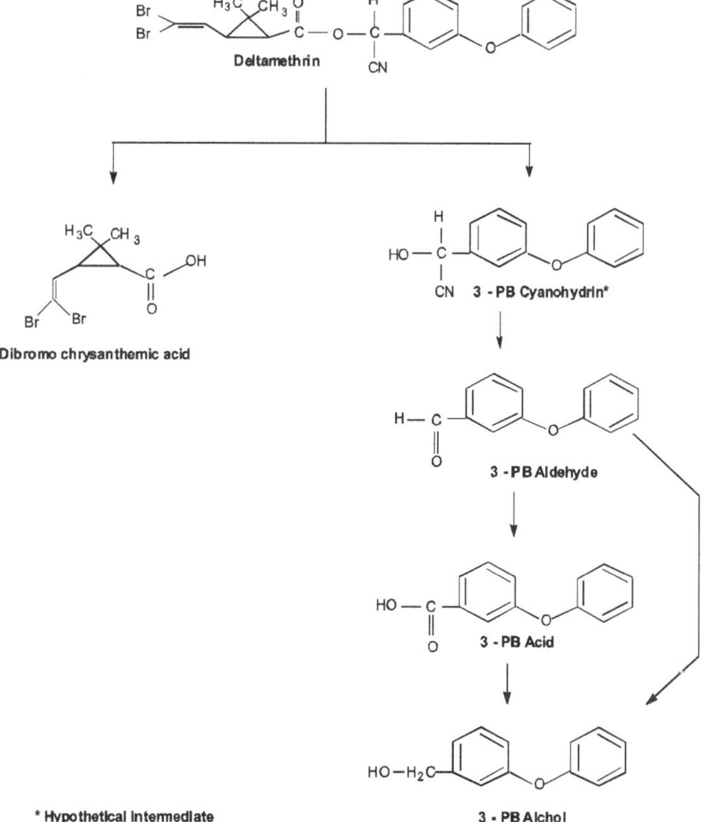

Figure 2.13: Metabolic pathways of deltamethrin in soil
(Source: Kathpal and Yadav, 2000)

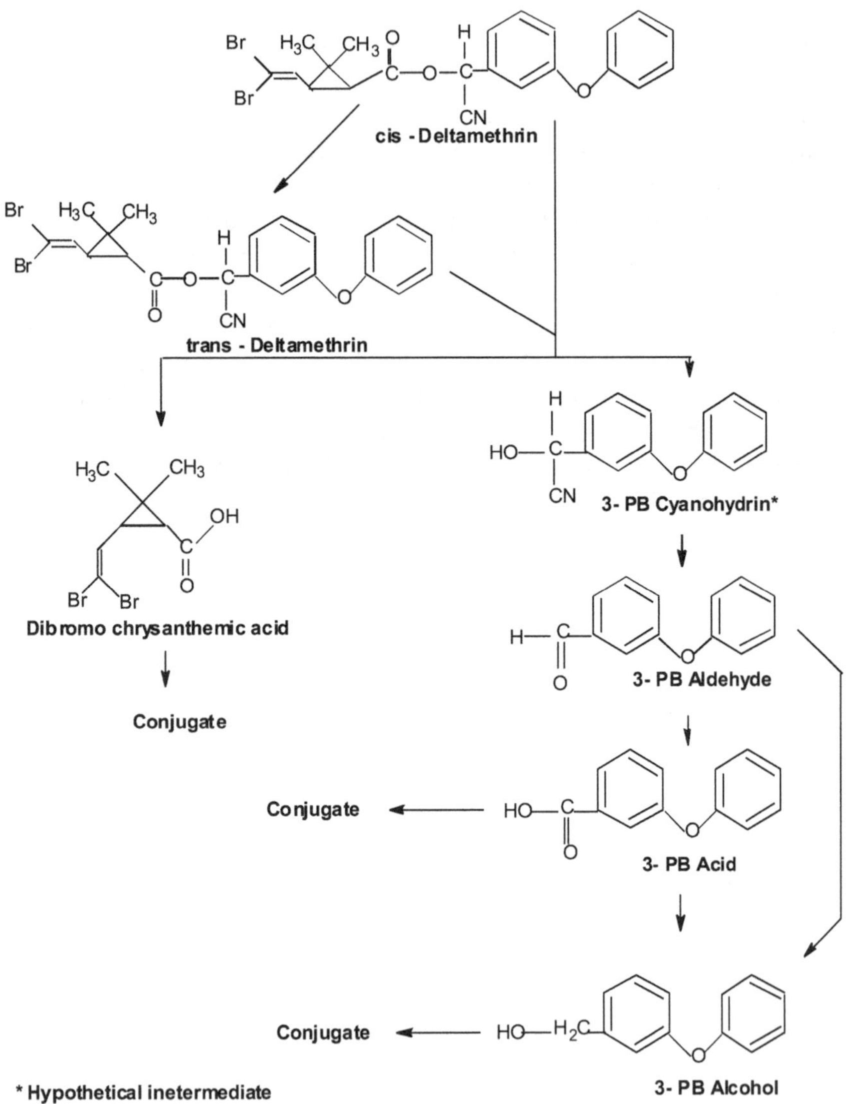

Figure 2.14: Metabolic pathways of deltamethrin in plants

(Source: Kathpal and Yadav, 2000)

2.6 Metabolism of Neonicotinoids

The neonicotinoids, the newest major class of insecticides, have outstanding potency and systemic action for crop protection against piercing-sucking pests, and they are highly effective for flea control on cats and dogs. They have been previously referred to as *nitro-quanidines*, *neonicotinyls, neonicotinoids, chloronicotines,* and more recently as the *chloronicotinyls*. These are a class of insecticides which act on the central nervous system of insects with lower toxicity to mammals. They are among the most widely used insecticides worldwide, but recently the uses of some members of this class have been restricted in some countries due to a possible connection to Colony Collapse Disorder which is the result of honey-bee populations collapsing. Their common names are acetamiprid, clothianidin, dinotefuran, imidacloprid, nitenpyram, thiacloprid, and thiamethoxam. Among these, the four neonicotinoids are acetamaprid (1), imidacloprid (2), nitenpyram (3) and thiamethoxam (4). They are potent broad spectrum insecticides having contact, stomach and systemic activity. The structures are given below and consist of substituted aromatic heterocyclic rings: 6-chloropyridinyl for acetamaprid, imidacloprid and nitenpyram and 3-chlorothiazole for thiamethoxam. Biotransformations involve some activation reactions but largely detoxification mechanisms. In contrast to nicotine, epibatidine, and other ammonium or iminium nicotinoids, which are mostly protonated at physiological pH, the neonicotinoids are not protonated and have an electronegative nitro or cyano pharmacophore. Agonist recognition by the nicotinic receptor involves cation-π interaction for nicotinoids in mammals and possibly a cationic subsite for interaction with the nitro or cyano substituent of neonicotinoids in insects.

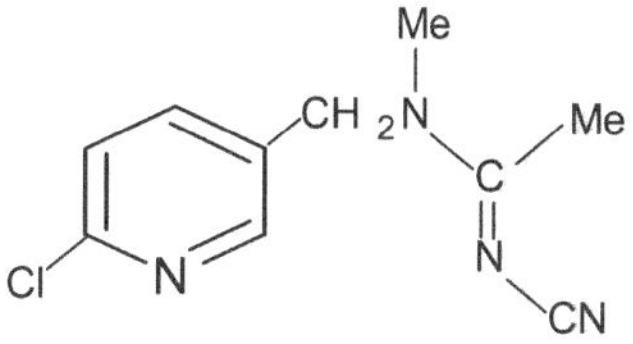

Acetamaprid (1)

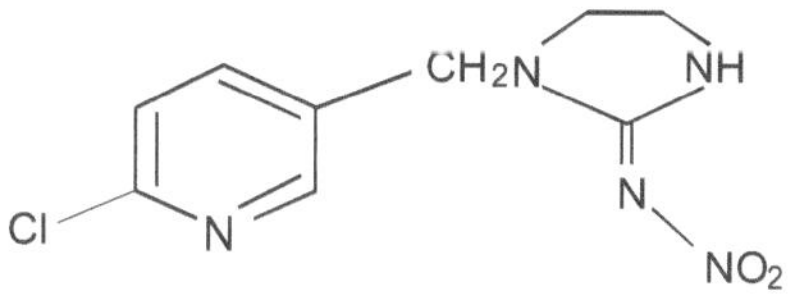

Imidacloprid (2)

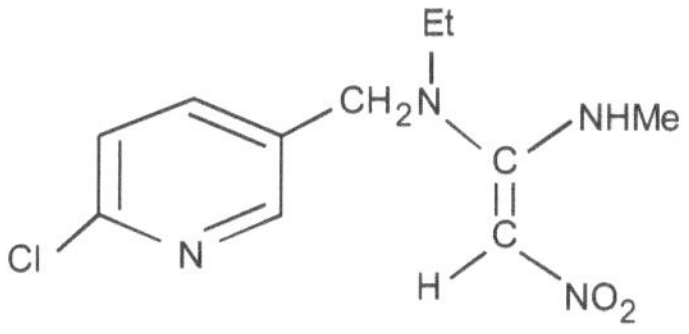

Nitenpyram (3)

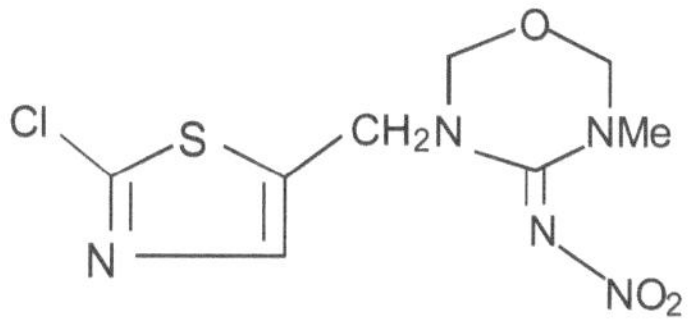

Thiamethoxam (4)

2.6.1 Metabolism of Acetamaprid

The metabpolic fate of [2, 6-^{14}C-pyridine] acetamaprid (Roberts and Hutson,1998) was investigated in eggplants and apples following application to foliage or fruit and cabbage plants after application to foliage and soil. The main component found in plant metabolism studies is acetamaprid. Small amounts of demethyl-acetamaprid (3), 6-chloronicotinic acid (6), 6-chloropicolyl alchol (7) and its conjugate with glucose (8) were produced. The metabolic pathways of acetamaprid in plants and soil are shown in Figure 2.15.

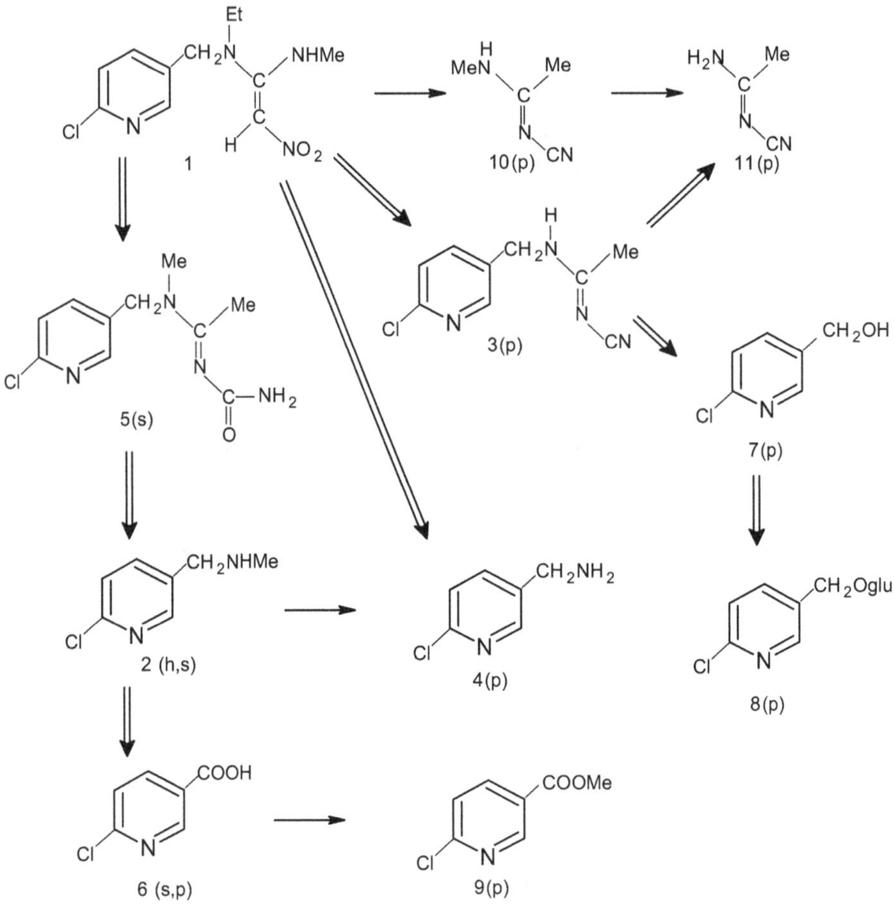

Figure 2.15: Metabolism of acetamaprid in soil(s) and plants (p)

(Source: Roberts and Hutson, 1998)

2.6.2 Metabolism of Imidaclorid

Imidacloprid is a systemic insecticide, having good root-systemic characteristics and notable contact and stomach action. It is used as a soil, seed or foliar treatment in cotton, rice cereals, peanuts, potatoes, vegetables, pome fruits, pecans and turf, for the control of sucking insects, soil insects, whiteflies, termites, turf insects and the Colorado potato beetle, with long residual control. Imidacloprid has no effect on mites or nematodes. Metabolism of imidacloprid in tobacco plant was studied in detail by Clark et al., 1998. Tobacco plants were grown in pots and the soil or foliage was treated with [^{14}C-methylene] imidacloprid. Leaves were sampled two weeks after the last spray application. The leaves were cured and made in to cigarettes and used for a smoking experiment. Tobacco leaves, cigarette butts and tobacco smoke were analysed using LSC and 2-D TLC methods. Imidacloprid in tobacco leaves was degraded to carbon dioxide during smoking. The only other components identified in smoke were uncharged parent imidacloprid, cyclic urea (2) and small amounts of carbon monooxide. The main compound in butts was parent imidacloprid and minor components were the cyclic guanidine (3), the cyclic urea (2), the olefin (12), the nitrosoguanidine (13) and the 4-hydroxy derivative (19).Metabolic pathway of imidacloprid in plants is shown in Figure 2.16. Metabolism of imidacloprid in rats is shown in Figure 2.17.

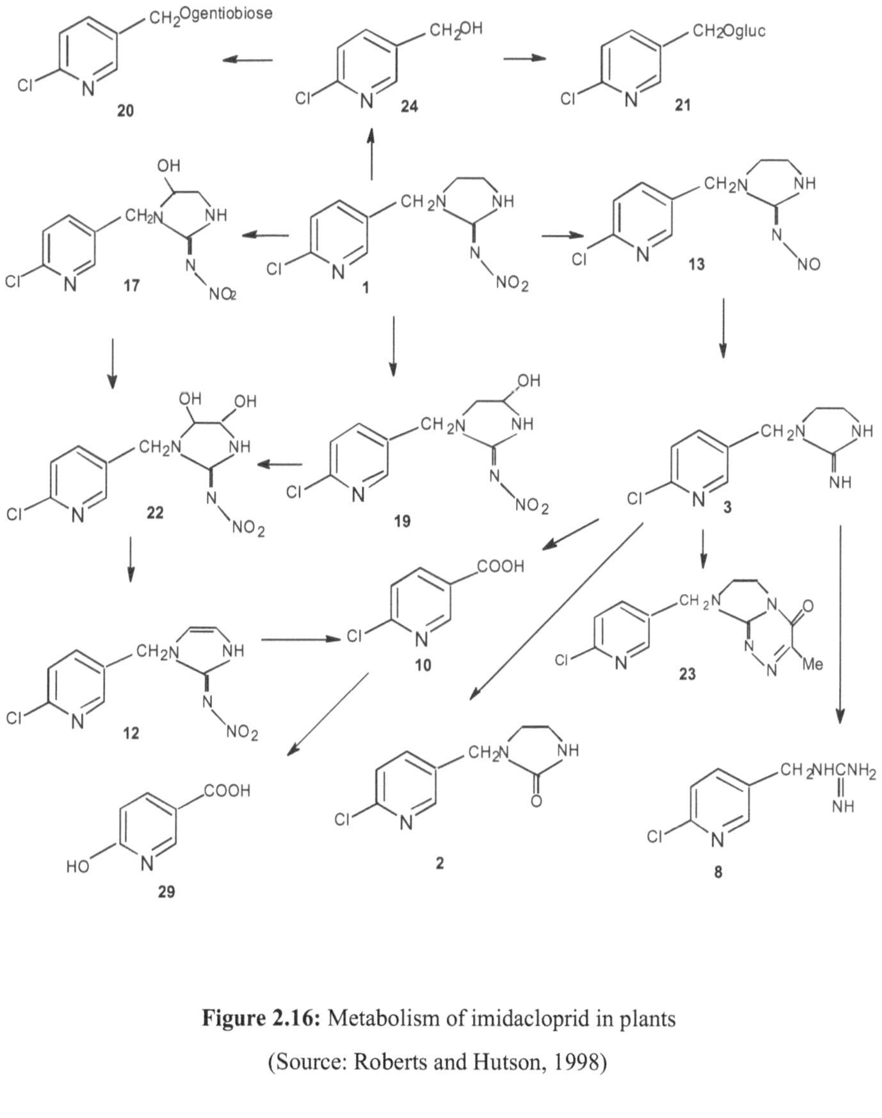

Figure 2.16: Metabolism of imidacloprid in plants

(Source: Roberts and Hutson, 1998)

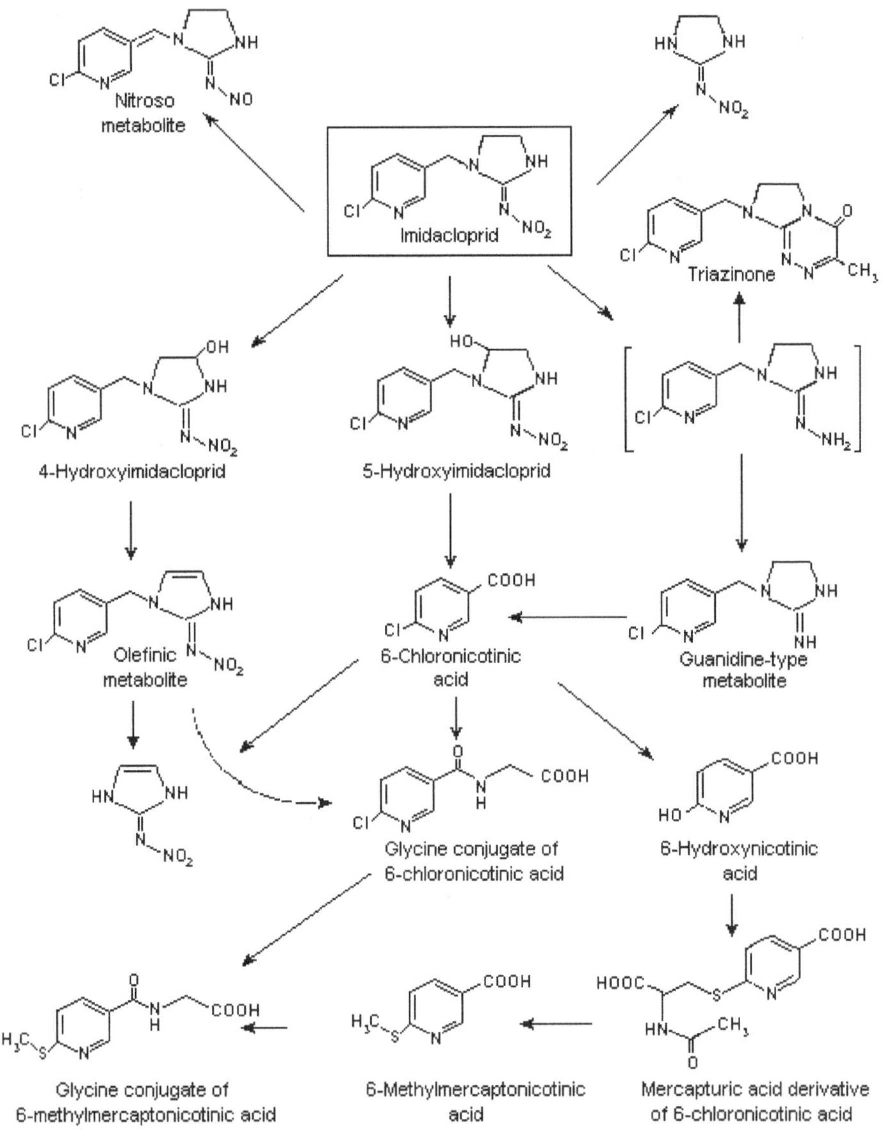

Figure 2.17: Metabolism of imidacloprid in rat

(Source: Robert and Hutson, 1998)

2.7 Metabolism of Herbicides

Among the herbicides, sulfonylureas are the important class in which the herbicides are used pre as well as post emergently. The sulfonylurea herbicides were discovered by Dr George Levitt at Du Pont in the mid 1970s during chemical scouting of the reaction of aryl sulfonyl isocyanates with aryl and aliphatic amines. The primary attributes of the sulfonylurea herbicides are their efficacy at low use rates, crop selectivity, negligible impact/adverse effects to the environment and safety to workers and consumers. The mode of action of sulfonylurea herbicides is inhibition of the biosynthesis of branched chain amino acids. Degradation and metabolism of sulfonylurea herbicides in water, soil, plants and animals involve multiple and complex transformation pathways. Although there is quantitative difference, primary degradation and metabolic pathways like cleavage of the sulfonylurea linkage, O-and N-dealkylation, aryl and aryl hydroxylation, ester hydrolysis and conjugation reactions with glutathione and carbohydrates are common among many sulfonylureas. Metabolic pathway of two recently used herbicides is explained.

2.7.1 Metabolism of Metsulfuron Methyl

Metsulfuron methyl is used for pre and post emergent herbicide for the control of annual and perenial broad leaved weeds of cereals, plantation crops and sugarcane. It undergoes degradation in water, soil, plants and animals by the combination of the following reactions: cleavage of the sulfonylurea linkage, hydroxylation of the phenyl ring and methyl carbon, O-demethylation and ester hydrolysis. The metabolic pathways of metsulfuron methyl are presented in Figure 2.18. Under aerobic conditions, the principal degradation metabolic pathway of [^{14}C-phenyl] metsulfuron methyl was the cleavage of the sulfonylurea linkage to yield **2, 4** and the acid sulfonamide **6** [2-(aminosulfonyl) benzoic acid]. Under anaerobic conditions, [^{14}C-triazine] metsulfuron methyl degraded to the primary metabolites, namely metsulfuron [2-(4-methoxy-6-methyl-1, 3, 5-triazin2-ylcarbamoylsulfamoyl)benzoic acid (7)] together with the Odemethylation products **8** [methyl 2-(4-hydroxy-6-methyl-1,3,5-triazin-2 ylcarbamoylsulfamoyl) benzoate] and **9** [2-(4-hydroxy-6-methyl-1, 3, 5 triazine-2-ylcarbamoylsulfamoyl)benzoic acid]. Compound **5** was also detected, probably formed as a hydrolysis product of the O-demethylated metsulfuron methyl **(8)** and/or metsulfuron **(9)**.

Hydroxylation in the phenyl ring of metsulfuron methyl [methyl 2(4-methoxy-6-methyl-1,3 ,5-triazin- 2-ylcarbamoylsulfamoyl) 4-hydroxybenzoate **(12)**] followed by carbohydrate conjugation was the major metabolic pathway in wheat and barley. Hydroxylation on the methyl carbon of the triazine moiety to yield methyl 2-(4-mcthoxy-6-hydroxymcthyl-triazin-2ylcarbamoylsulf amoylbenzoate **(13)** was observed as a minor pathway. The cleavage of the sulfonylurea linkage to form 2, 4, and 6 from the ^{14}C-phenyl]-moiety and 4-methoxy-6-hydroxymethyl-2-amino-1, 3, 5 triazine **(14)** from the ^{14}C—triazine] moiety (Anderson et al., 1989) was also observed as a minor metabolic pathway. Minimal transfer of metsulfuron methyl residues to milk, egg, organs, and tissues in goat and chicken was reported (Charlton and Bookhart, 1995).

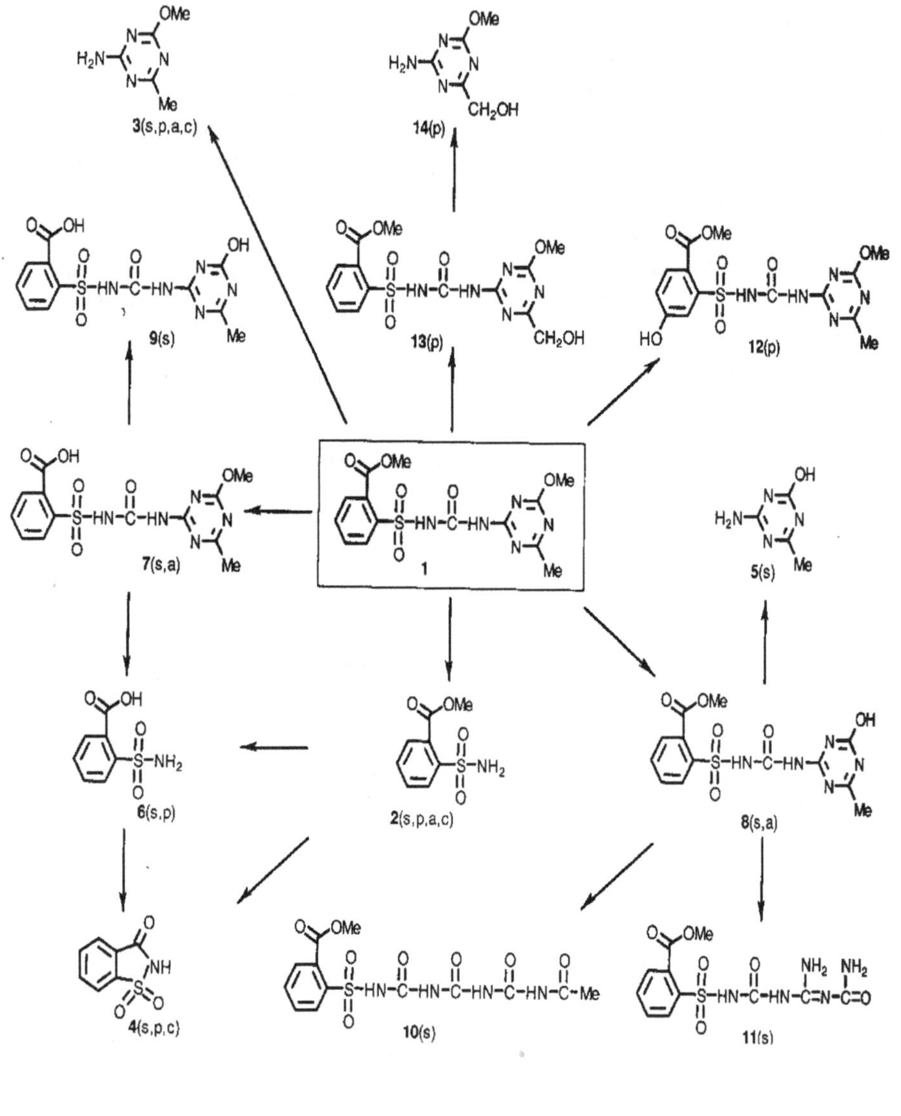

Figure 2.18: Metabolic pathway of metsulfuron methyl in soil(s), plant (p) and animal (a)

(Source: Roberts and Hutson, 1998)

2.7.2 Metabolism of Tribenuron Methyl

Tribenuron methyl degrades in acidic solution and soil primarily through cleavage of the sulfonylurea linkage. Hydroxylation of the phenyl ring followed by rapid carbohydrate conjugation is the primary metabolic pathway in wheat. Other metabolic reactions include O-demethylation of the methoxy group in the triazine ring and hydrolysis of the parent methyl ester. N-demethylation of the tribenuron methyl to yield metsulfuron methyl occurred in both animal and plant systems, but has not been observed in soils. Tribenuron methyl is unique among commercial sulfonylureas in having N-methyl substituent on the sulfonylurea linkage. This feature makes this modified linkage much more susceptible to hydrolytic cleavage. This increased susceptibility of this herbicide to hydrolysis in water and soil allows its use with minimal re-cropping restrictions.The degradation and metabolic pathways of tribenuron methyl are presented in Figure 2.19.

Tribenuron methyl degraded rapidly in an acidic soil at 25°C under laboratory aerobic incubation conditions (Rapisarda and Scott, 1986). The primary degradation pathway is the cleavage of the sulfonylurea linkage to yield **2**, **3** and **4**. Further degradation of 4 involved 0- or N-demethylation yielding 4-hydroxy-6-methyl2-methylamino-1,3,5-triazine **(5)** and 4-methoxy-6-methyl-2-amino-1,3,5triazine **(6)**, respectively.Although chemical hydrolysis remained an important soil degradation pathway, yielding **2, 3, 4, 5, 6 and 7** (4-hydroxy-6-methyl-2-amino-1, 3, 5-triazine), soil microorganisms also played a major role under anaerobic conditions. Microbially mediated decomposition of tribenuron methyl under anaerobic conditions also included the O-demethylation of tribenuron methyl to yield methyl 2 - [4- hydroxy -6-methyl-1, 3, S-triazine - 2 - yl(methyl)carbamoylsulfamoyl]benzoate **(8).** Hydrolysis of the methyl ester of tribenuron methyl yielded tribenuron {2 -[4-methoxy-6-methyl-1,3 ,S-triazine-2 -yl(methyl)carbamoy 1sulfamoyl]benzoic acid **(9)**}.Compound **10**{2-[4-hydroxy-6-methyl1, 3, S-triazine-2-yl(methyl)carbamoylsulfamoyl]benzoic acid} was formed by the hydrolysis of **8** or *via* the O-demethylation of **9.** Acid benzenesulfonamide (11) was also observed, probably as the hydrolysis product of **9** and/or methyl benzenesulfonamide **(2).** N-demethylation of tribenuron methyl to form another active sulfonylurea herbicide, metsulfuron methyl {methyl 2-[4-methoxy-6-methy 1-1,3 ,S-triazine- 2 -ylcarbamoylsulfamoy1]benzoate **(12)**} was the major pathway. Metsulfuron methyl was rapidly metabolised, primarily through the hydroxylation of the phenyl ring {(methyl 2-[4-methoxy-6-methyl-1,3,S-triazin-2-yl-carbamoylsulfamoyl]4-hydroxybenzoate **(13)**)followed by carbohydrate conjugation. Hydrolysis of the sulfonylurea linkage of tribenuron methyl, metsulfuron methyl (12), and hydroxylated metsulfuron methyl **(13)** were also important metabolic/degradation pathways. A major metabolic pathway in animals was N-demethylation to yield metsulfuron methyl **(12)** which was excreted mainly unchanged. Compounds **3, 4, 5, 9** and **11** were the predominant metabolites recovered in the rats, while **3** and **7** were significant products recovered from goat excreta. Other minor metabolites detected from both animal species included **2, 6, 13** and **15.**

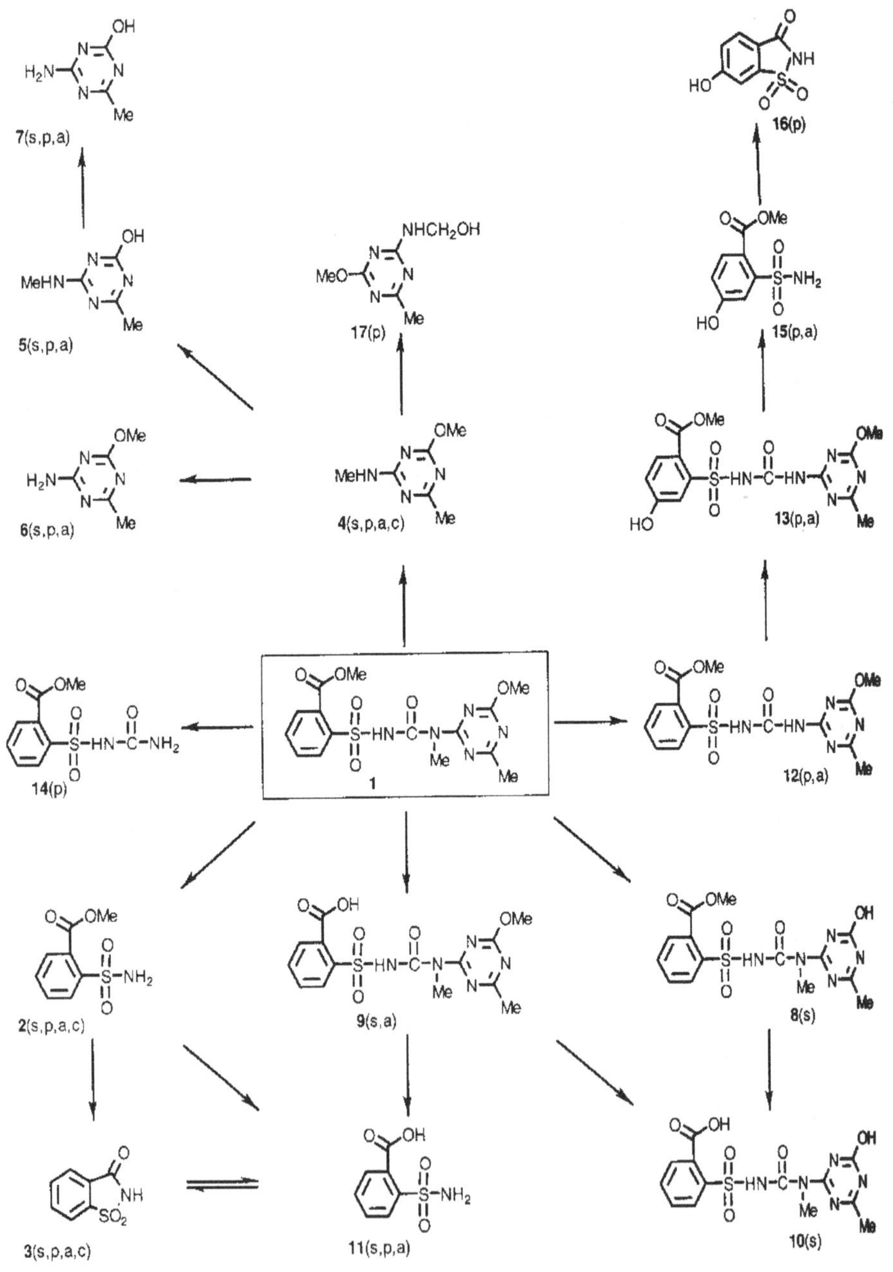

Figure 2.19: Metabolic pathway of tribenuron methyl in soil(s), plant (p) and animal (a)

(Source: Roberts and Hutson, 1998)

References

Ahmad, S., Brattsten, L.B., Mullian, C.A. and Yu, S.J., 1986. Enzymes involved in the metabolism of allelochemicals. In: *Molecular aspects of insect-plant associations* (eds. Brattsten, L.B. and Ahmad, S.), Plenum Press, New York, USA.

Aizawa, H., 1982. Metabolic maps of pesticides. In: *Ecotoxicology and environmental quality series*. Academic Press, Inc. London, UK.

Anderson, J.J., Priester, T.M. and Shalaby, L.M., 1989. Metabolism of metsulfuron-methyl in wheat and barley. *J. Agric. Fd. Chem.* 37: 1429-1434.

Andrawes, N.R., Dorough, H.W., and Lindquist, D.A., 1967.Degradation and elimination of temik in rats. *J. Econ. Entomol.* 60 (4): 979-987.

Bartely,W. J. Nathan, R.,Andrawes, N.R.Chancey, E.L., Bagley,W.P., and Spurr, H.W., 1970. The metabolism of Temik aldicarb pesticide [2-methyl-2(methylthio) propinoaldehyde O-(methylcarbamoyl) oxime] in the cotton plant. *J.Agric.Fd.Chem.* 18: 446-453.

Bromiow, R.H., Briggs, G.G., Williams, M.R., Smelt, J.H., Tuinstra, l.G.M., and Tragg, W.A.,1986. The role of ferrous ions in the rapid degradation of oxamyl, methomyl and aldicarb in anaerobic soils. *Pestic. Sci.* 17(5): 535-547.

Charlton, R.R., and Bookhart III, S.W., (1995) Metabolism of [^{14}C] metasulfuron methyl in laying hens. DuPont Internal Report, AMR 3554-95.

Clark, A.G., and Shamaan, N.A., 1984. Evidence that DDT-dehydrochlorinase from the house fly is a glutathione S-transferase. *Pestic. Biochem. Physiol.,* 22 (3): 249-261.

Clark, A.G., Shamaan, N.A., Dauterman, W.C., and Hayaoka, T., 1984. Characterization of multiple glutathione transferases from the house fly, *Musca domestica* (L.). *Pestic.Biochem.Physiol.*22: 51-59.

Clark, T., Kaussmann, E., and Schepers, G., 1998.The fate of imidacloprid in tobacco smoke of cigarettes made from imidacloprid-treated tobacco. *Pestic. Sci.* 52:119-125.

Coppedge, J.R., Lindquist, D.A., Bull, D.L., and Dorough, H.W., 1967. Fate of 2-methyl-2-(methlthio) proppioonaldehyde O-(methylcarbomyl)oxime (Temik*) in cotton plants and soil. J.Agric.Fd.Chem.*15:902 910.

Croucher, A., Hutson, D.H., and Stoydin, G.,1985. Excretion and residues of the pyrethroid insecticide cypermethrin in lactating cows. *Pestic.Sci.* 16:287-301.

Dauterman, W.C., 1985. Insect metabolism: Extramicrosomal. In: *Comprehensive insect physiology* (eds. Kerkut, G.A. and Gilbert, L.I.), Pergamon Press, New York, USA: 713.

Dorough, H.W., and Ivie, J.E., 1968. Temik-35S metabolism in a lactating cow. *J. Agric. Fd. Chem.,* 16 (3): 460–464.

Hassal, K.A., 1990. *The Biochemistry and Uses of Pesticides. Structure, Metabolism, Mode of Action and Uses in Crop Protection.* Pub. VCH Suite, New York, USA.

Hodgson, E., and Levi, P.E., 2001.Metabolism of pesticides. In: *Handbook of pesticide toxicology,* 2nd edn. (ed. Krieger, R.I.), Vol.2, Academic Press, San Diego,USA:531.

Hodgson, E.,1985. Microsomal mono-oxygenases. In: *Comprehensive insect physiology, biochemistry and physiology* (eds. Kerkut, G.A. and Gilbert, L.I.), Pergamon Press, New York, USA, 11: 225.

Hutson, D.H., and Stoydin, G., 1987. Excretion and residues of the pyrethroid insecticide cypermethrin in laying hens. *Pestic. Sci.* 18:157.

Kathpal, T.S. and Yadav, P.R., 2000. Insecticide metabolism in insects, animal and plants. In: Pesticide residue analysis (eds.Yadav,P.R., Kathpal,T.S. and Rohilla, H.R.), Department of Entomology, CCSHAU, Hisar: 127-138.

Knaak, J.B., Tallant, M.J., and Sullivan, L. J., 1966. The metabolism of 2-methyl-2-(methylthio) propinoaldehyde O-(methylcarbamoyl) oxime in the rat. *J.Agric.Fd.Chem.*14:573-578.

Kuhr, R. J., 1970. Metabolism of carbamate insecticide chemicals in plants and insects. *J. Agric. Fd. Chem.* 18 (6): 1023–1030.

Kuhr, R.J., 1971. The formation and importance of carbamate insecticide metabolites as terminal residues. *Pure and Appl. Chem. Suppl.*:199-220.

Kuhr, R.J., and Dorough, H.W., 1976. *Carbamate Insecticides: Chemistry, Biochemistry and Toxicology.* CRC Press Cleveland. Ohio, USA.

Li, X., Schuler, M.A., and Berenbaum, M.R., 2007. Molecular mechanisms of metabolic resistance to synthetic and natural xenobiotics. *Annu. Rev. Entomol.* 52: 231.

Matsmura, F., 1985.*Toxicology of Insecticides*.2nd ed. Plenum Press, New York, USA.

Menzie, C.M., 1974. Metabolism of pesticides. An update : US Department of the Interior Fish and Wildlife Service,Special Scientific Repot-Wildlife No. 184, Washington, DC,USA.

Metcalf, R. L., Fukuto, T. R., Collins, C., Borck, K., Burk, J., Reynolds, H. T., Osman, M. F.,1966. Metabolism of 2-methyl-2-(methylthio) propionaldehyde O-(Methylcarbamo yl) oxime in plant and insect. *J. Agric. Fd. Chem.*, 14 (6): 579–584.

Nakatsugawa, T., and Morelli, M.A., 1976. Microsomal oxidation and insecticide metabolism. In: *Insecticide biochemistry and physiology (*ed. Wilkinson, C.F.), Plenum Press, New York, USA: 61.

Ou, T., Gancarz, D. H., Wheeler, W.B., Roa, P.S.C. and Davison, J.M., 1982. Influence of soil temperature and soil moisture on degradation of metabolism of carbofuran in soils.*J. Environ. Qual.* 11: 293.

Pekas, J. C., 1979 Further metabolism of naphthylN-methylcarbamate (carbaryl) by the intestine. *Pestic. Biochem. Physiol.* 11: 166.

Rapisarda, C., and Scott, M.T., 1986. Aerobic soil metabolism of [^{14}C-*Triazine*]*DPX* L5300. DuPont Internal Report, AMR 360-85, Rev. 2.

Roberts, T.R., and Hutson, D.H., 1998. *Metabolic Pathways of Agrochemicals,* Part 2: *Insecticides and Fungicides.* Cambridge, The Royal Society of Chemistry, London.

Roberts, T.R., and Hutson, D.H., 1998. *Metabolic Pathways of Agrochemicals,* Part 1: *Herbicides and Plant Growth Regulators.* Cambridge, The Royal Society of Chemistry, London.

Robinson, J., 1969. *Can. Med. Assoc. J.* 100: 80.

Walker, C. H., 1975. In: *Organochlorine insecticides*: Persistent Organic Pollutants (ed. Moriarty, F.), Academic Press, New York, USA: 73.

www.inchem.org/documents/jmpr/jmpmono/v00pr06.htm. Pesticide residues in food 2000: Fenitrothion First draft prepared by U. Mueller Chemicals and Non-Prescription Medicines Branch Therapeutic Goods Administration, Canberra, ACT, Australia

Yang, R.S.H., 1976. Enzymatic conjugation and insect metabolism. In: *Insecticide biochemistry and physiology* (ed.Wilkinson, C.F.), Plenum Press, New York, USA:177.

Yu, S.J., 2004.Detoxification mechanisms in insects. In: *Encyclopedia of entomology* (ed. Capinera, J.L.), Vol.1, Kluwer Academic Publishers, Boston, USA: 686.

Yu, S.J., 2008.The toxicity and biochemistry of insecticides. CRC Press, Taylor & Francis Group, New York, USA.

Chapter 3

Sampling, Extraction and Clean-up

After having performed their function, pesticides in the parent form or in some converted form reach the soil, the ultimate sink for these moieties which are collectively termed as residues. According to FAO 1986, pesticide residue is defined as any substance or mixture of substances in food, agricultural commodities or animal feed resulting from the use of pesticides and includes any specified derivatives such as degradation products, metabolism, reaction products and impurities that are considered to be of toxicological significance. From the soil they move into other segments of the environment. The pesticide residues are absorbed by plants and enter the food chain, where they get concentrated in animal fat. Some residues are carried away by wind in the vapor form, which enables them to be transported to long distances from the source of application and then get collected by dust and rain before they are deposited on the ground again. Pesticide residues also move to aquatic environments through runoff, leaching, and atmosphere disposition.

Pesticide residues occur in agricultural commodities because of:

- Contamination of crops or animals exposed to chemicals in the environment.
- Intentional use of pesticides for protection of growing crops or stored products or of animals.
- Unintentional exposure to pesticides such as would occur in crops, grown in soil treated previously or contaminated by foliar treatment of other crops grown earlier in the rotation.
- Unintentional accumulation of pesticide residues in animals exposed to chemicals.

Residues of the pesticides are present in micro quantities in the matrix, hence involves a complicated procedure involving many steps for their analysis. Analysis does not depend on the high cost equipments like gas liquid chromatograph (GC) and high performance liquid chromatograph (HPLC) but is based on series of cumulative operations like sampling strategy, storage of the sample, sample treatment, sample extraction and clean-up techniques . All these operations are probable sources of inaccuracy and imprecision that can inadvertently be introduced into the entire analytical procedure.

3.1 Steps involved for Carrying out Residue Analysis

The basic steps involved in pesticide residue methodology are:

3.1.1 Sampling
3.1.2 Design of experiment
3.1.3 Sample preparation (preparation of laboratory samples)
3.1.4 Extraction
3.1.5 Clean-up
3.1.6 Estimation

3.1.1 Sampling

Sampling can be defined as the procedure or step adopted to obtain a representative quantity from the large consignment, so that the selected representative quantity can be handled conveniently. The residue contents of agricultural commodities are determined for getting such data to support a petition for establishing tolerance for a pesticide chemical or for finding out whether or not a certain lot of a commodity contains an illegal, above-tolerance residue level. Sampling is an important consideration in either case. In establishing reliable sampling methods or procedures, much thought, care, and ingenuity are needed to acquire a representative laboratory sample for residue analysis.

3.1.1.1 Sampling Strategies

The main objective in any sampling strategy is to obtain a representative portion of the sample. This requires a detailed plan of how to carry out the sampling. Therefore planning of the sampling strategy is an important part of the overall analytical procedure as the consequences of a poorly defined sampling strategy, as well as costing both time and money, could well lead to getting the wrong answer. The necessary criteria for carrying out an effective sampling strategy for environmental analysis are given below:

3.1.1.2 Criterion for Effective Sampling

For effective sampling, specific objectives are to be set first. Proper arrangements are made to obtain samples from the sites. Special equipments for sampling and container for storing the samples are also made ready. Prior planning regarding number of samples to be collected, approach like random, systematic, judgemental, procurement of control sample (where no pesticide is applied) is also quite essential.

3.1.2 Design of Experiment

3.1.2.1 Comparative Planning

In crop residue studies, the design of the experiment is of critical and paramount importance. For this reason, all such studies should evolve from cooperative planning on the part of

(i) the entomologist, plant pathologist or other crop scientist
(ii) the agronomist
(iii) the analytical chemist

3.1.2.2 Formulation

Major trials should be carried out with proposed commercial formulations. It is meaningless to carry out this work with laboratory prepared formulations since the fate of residues may be influenced by the nature of formulations. It is preferable to make the application with commercial equipment in a manner analogous to that used by farmers, but the greatest care should be exercised to see that the application is uniform throughout.

3.1.2.3 Maximum Residue Limit

Since one of the objectives of residue studies is to provide the basis for the estimation of maximum residue levels, the design of the experiments should be directed to the determination and evaluation of the conditions and factors that lead to the highest residue levels following recommended use patterns.

3.1.2.4 Type / Variety of Crops

The trials to determine residue levels should be repeated on different varieties, different stages of growth of typical periods of the year and under different agricultural regimes to determine residue levels under various agro-climatic conditions.

3.1.2.5 Selection of Sites

Trials should be carried out in major areas of cultivation or production and should cover the range of representative conditions (climate, seasons, soil, cropping system, farming etc.)

3.1.2.6 Sampling Considerations

3.1.2.6.1 Characteristics of the Sample

Sample should be truly representative. It should be accurate, valid, representative and variable. Accuracy depends on the field from where it is collected. Valid sample is one which, when selected ensures equal chances for each unit of the material in the population being sampled. A representative sample is not only a random sample but also that the proportion of each type of the substrate composition is identical to that of the gross sample from which it is selected.

3.1.2.6.2 Plot Size

Residue data should not be generated from plots that are too small to be representative. The size of the individual plot may vary from crop to crop but should be large enough to allow application of the pesticide in an accurate and realistic manner. A sufficient buffer zone should be left between plots to prevent cross-contamination.

3.1.2.6.3 Dosage Rate

At least two dosage rates should be included in each residue trial design, the maximum rate, which is likely to be recommended, and another rate preferably double the recommended rate.

3.1.2.6.4 Control Plots

A control plot for the supply of untreated samples is necessary. The control plots should be large enough to satisfy these requirements and should be located close enough to secure identical growing and climatic conditions. Control samples are needed:

(i) to ascertain that no artifact in the crop derived from local conditions could give rise to interference in the analysis

(ii) to establish the recovery level of the pesticide from the crop or soil by analytical method

(iii) in the case of new crop, to investigate the storage stability of any residue

3.1.2.6.5 Replications of the Sample

Larger the sample replicates, greater is the validity of the results but there is a limit to numbers of replicates. From statistical consideration, each field plot should be replicated at least three times and there should be at least three sample replicates per plot.

3.1.2.6.6 Time Element

Residue status at different time intervals i.e. at harvest, post harvest and terminal stage shall differ. It may be few days in plants but in soil, it may be months and years also. For working out degradation or persistence behaviour, timing of sampling collection is very critical.

3.1.2.6.7 Type of Commodity

All the parts of plant like leaf, fruit, flower and seed as well as their substrates require due attention in collecting samples. Different samples are collected in a different manner.

3.1.2.6.8 Storage of Samples

Sometimes samples cannot be analysed immediately upon receipt, then some form of sample storage is almost always required. The concern with the storage of samples is that losses can occur, due to adsorption to the storage vessel walls, or that potential contaminants can enter the sample, from desorption or leaching from the storage vessels. These problems can all lead to the analyst getting the wrong answer. The goal therefore is to store samples for the shortest possible time interval between sampling and analysis. Indeed in some instances where analytes are known to be unstable or volatile it may be necessary to perform the analysis immediately upon receipt or not at all. The nature and type of storage container is also important. Light sensitive samples should be stored in a brown glass container to prevent photochemical degradation. Volatile type of samples species is stored in a well-sealed container. In most cases, the use of glass containers is recommended as there is little opportunity for contamination because of container. In case of liquid samples, it is better to completely fill the storage container rather than leave a significant headspace above the sample. This acts to reduce any oxidation that may occur. In addition to glass containers, polyethylene or poly tetrafluoro ethylene (PTFE) containers are appropriate to use for solid samples. Plastic containers are not recommended for aqueous samples as can cause problems at later stages of the analysis, e.g. phthalates.

During storage, pesticides in samples of plant etc. deteriorate. This is particularly with organophosphates. Barcelo and Alpendurada (1996) highlighted the problems associated with the storage and preservation of polar pesticides in water samples. It has been identified that in some instances the degradation products of pesticides are more stable than the parent compounds so that emphasis, in terms of the analysis, should also be aimed at the degradation products. Chiron et al., (1993) reported that some carbamate pesticides (methiocarb sulfone, methiocarb sulfoxide and 3-ketocarbofuran) are stable in water samples for up to 20 days whereas carbaryl losses can be as high as 90% in one day (Marcos et al., 1995). Hence the collected samples should be analysed without loss of time. If quick analysis is not possible due to some unavoidable circumstances, samples should be stored in deep freezer (-20-40^{0}C). Storage under these conditions reduces most enzymatic and oxidative reactions. It is suggested that storage period be kept minimum and suitable technique of addition of pesticide arrestants of fortification should be adopted to check the deterioration in the sample.

3.1.2.6.9 Maintenance of Record

In order to avoid mixing, all samples collected should be properly labeled. This will help in finding out the history of the sample. Transference of minute amount of pesticide residue from the matrix into the solvent is termed as extraction and the efficiency of extraction method rely on the establishment of equilibrium between the toxicant, concentration in the recovered solvent and that remaining in the matrix.

3.1.2.6.10 Sampling Procedures for Different Environmental Components

Water

For rivers, the primary sampling point is in the surface water layer (0-5 cm from the surface) at the centre of the main flow. However, the top 1-2 cm of this surface layer should be avoided so as not to collect floating dust, oil, etc. In addition, further samples can be collected through the full depth of the water column if required to meet the specific purpose of the study. For underground water, the sampling site or sites will be selected after taking into consideration such factors as water flow and geological structure and also site conditions such as factories or land use, and avoiding bias so as to be able to understand the whole area's underground water. Collect 3 litres of water to represent 1 sample from 1 water source, and transport the sample to the laboratory and keep at 4°C or at lower temperature. Flow chart for sampling procedures for water is shown in Figure 3.1.

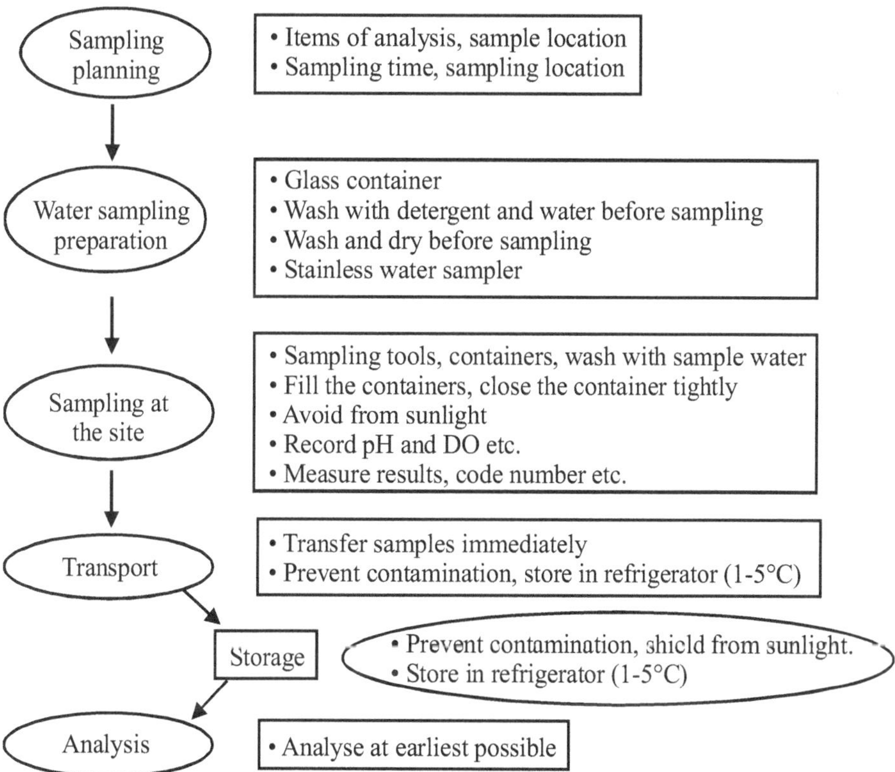

Figure 3.1: Water sampling procedure

Soil

The sampling of soil is typically done to detect pesticide residues or to routinely monitor environmental samples (Sharma, 2007). Soil samples should be taken from growing fields in the grid pattern uniformly distributed, so that each area of the field is sampled. A 3 × 3 grid with 9 total sample portions is suggested for smaller fields, with 4 × 4 (16 sample portions) for the medium-sized fields, and 5 × 5 and even larger grids are used for very large fields. Each sample site represents 1 portion of the total sample, and at each site, 2 soil plugs about 15-cm deep and 3 to 5 cm in diameter are to be taken. The 2 plugs, when combined, become sample portion of that sample site (Figure 3. 2). Another common soil sampling method for a field or other area is to take '5' portions in a 'Z' pattern. An example of 3 × 3 sampling grid is designed by X pattern sampling.

Sampling tools include soil augers. Place each portion of the soil sample into a separate glass jar covered with aluminium foil. It is recommended to chill soil samples to 4°C for transport to the laboratory. The glass jars for collecting soil and water samples should be rinsed thoroughly with acetone/methanol and dried.

Z-pattern soil sampling **X-pattern soil sampling**

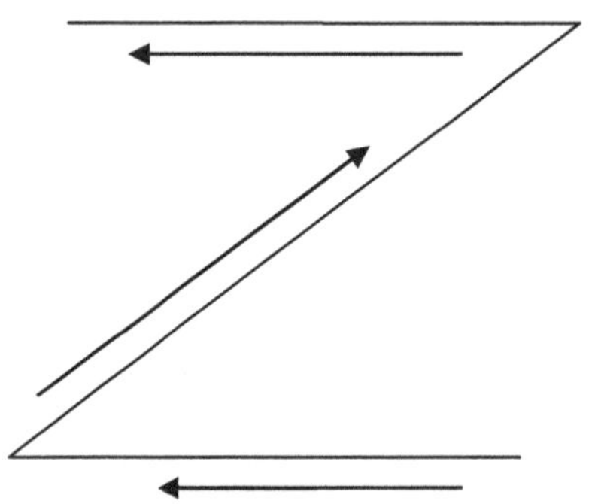

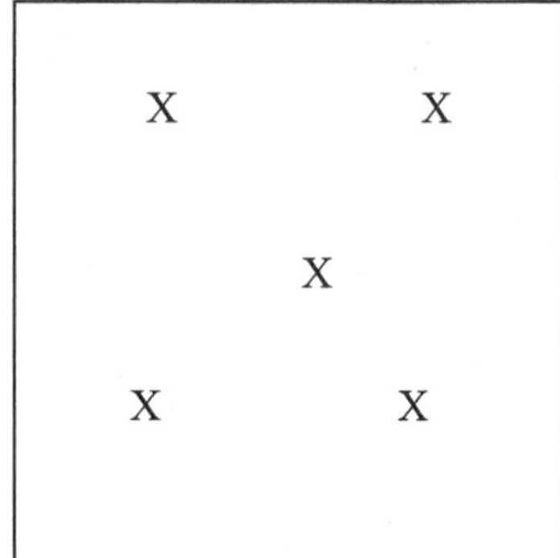

Figure 3.2: Soil sampling pattern
(Source: Sharma, 2007)

Collection of the Gross Sample

In most cases, it is not practical to collect the entire crop from a test plot. Therefore, it is generally necessary to devise means of selecting a sample referred to as gross sample or field sample that when reduced and analysed will give a residue result which for all practical purposes, will be equal to the actual average residue level on the entire crop in the plot. In general, the gross sample must be representative, of the treated plot. The amount of different commodities required to constitute a satisfactory sample obviously vary according to the commodity. The amounts indicated in Table 3.1 have been found satisfactory by *Codex Committee* for pesticide residue analysis.

3.1.2.6.11 Sampling Procedures for Different Food Commodities

Vegetables and Fruits

As far as possible, a sample of a fruit or a vegetable must be taken directly from the field or grower at the time of, or shortly after the harvest for monitoring studies. The history of the pesticides applied to a given field is extremely important, and such information is vital to the analysis. Superimpose an imaginary grid on the field dividing it into approximately 100 areas, and randomly select 10 of these areas to form a representative sample of the field. Collect 0.5-1 kg portion from each of these 10 areas and combine them to form the sample. In case of leafy vegetables (cabbage, lettuce etc.), the portions collected from a growing field should be representative of the local commercial practice. If the fruit and vegetables samples are to be taken at transport (or) storage, collect 10 samples, one kg each, randomly from the lot. It is recommended that a minimum total sample consisting of the recommended number of portions, depending on the lot size, should be collected.

Grains

Grains are never sampled at the grower or primary producer. If such samples are to be taken, the sampler can use the above sampling procedure. But sampling of grains during transport or shipments and storage is important, because residues on grains are typically due to post-harvest use of fumigants. A sample size of 5 kg should be taken from 5 locations.

Milk

Fluid milk, like grains, should never be sampled at primary producer. Since fluid milk is relatively homogeneous, a total sample size of 2 litres is usually sufficient from the producer or each tank. Sample portions need not be packed separately and can be combined in one sample container. Glass jars are most commonly recommended. Where pouched milk is to be sampled, a minimum of three 500 ml pouches of similar nature need to be taken.

Animal Feeds

Animal feeds including hay, silage, grains, byproducts and commercial feed rations, are normally sampled, only when there is a reason to believe that a pesticide problem exists in food. Therefore, sampling procedure is typically selective, not random. If the feed is the bulk, collect 10 x 1 kg portions to comprise one sample.

Table 3.1: Minimum sample sizes for fruits and vegetables required for sampling

Type of food	Minimum amount of sample to be collected
Small or light foods with unit weight of 25g or less (e.g. Berries, peas)	1 kg
Medium-sized foods with unit weight usually between 25 and 250g (e.g. Apples, oranges, carrots, potatoes)	1 kg (at least 10 fruits)
Large-sized food with unit weight more than 250g (e.g. Cabbage, cauliflower, melons, cucumbers etc.)	2 kg (at least 5 fruits)
Dairy products like, butter, cream, milk and cheese	0.5kg
Oils and fats (Cottonseed)	0.5kg
Spices (Coriander, cumin)	0.25kg
Cereals (Wheat, rice)	1.0kg
Animal feed (Cotton seed cake)	1.0kg

(Source: Handa et al., 1999)

3.1.2.6.12 Sampling Procedures for Processed Foods for Monitoring Programmes

The key consideration is the packaging size, because the degree of processing can vary considerably. The foods to be sampled need not be divided into groups as is done for raw and partially processed foods. When the samples of processed food are taken at the processor, the sampler has a unique opportunity to obtain information about the sample that usually is not available. Codex has recommended sampling schedule for all the processed packaged foods, and this is given in Table 3.2.

Table 3.2: Codex recommendation for the sampling

Number of retail units in the lot	Minimum number of retail units to be sampled
1 - 25	1
26 - 100	5
101 - 250	10
Over 250	15

(Source: Sharma, 2007)

3.1.3 Sample Preparation

The selection of the gross sample may result in the accumulation of 1-5 kg of material that cannot be simply mixed and quartered as required to produce a laboratory sample which is truly representative of gross sample and in turn, of the residue level in the experimental plot. It is necessary at this stage to do any cleaning, pruning and trimming required. Select the crop components of interest, reduce the size of the component parts, mix, subdivide, and systematically reduce the sample size to 250 g without altering the average residue level of the sample. Special problems exists in the sampling of natural waters due to their homogeneity and relatively little has been done in the area of sampling of rain water for pesticide residue analysis. Efforts should be made for developing procedures for sampling of air, rainwater, natural water, and collaboration between statisticians and analysts are required to develop improved guidelines for sampling procedures.

3.1.4 Extraction

Once a valid, representative sub-sample has been selected for residue analysis, it is processed for isolation of pesticide or its metabolites having toxicological significance from the surrounding biological environment. Extraction must be adequate to yield quantitative removal of the toxicant.

3.1.4.1 Selection of Extraction Techniques

3.1.4.1.1 Before selecting the technique, one should be aware of:

a) Nature of matrix, whether soil, plant material or water etc.

b) The analyte's chemical group, its metabolites and whether these are thermally stable or not.

c) Safety, (method to be adopted should not be hazardous).

d) Co-extractives (presence of excessive co-extractives requires drastic clean-up measures leading to loss of active principle)

e) One should ensure that the analyte /metabolite are fairly soluble in the selected single solvent or mixture of solvents.

3.1.4.1.2 Characteristics of Solvents

The selection of solvents should be accordingly as 'Like dissolves like'. For non polar analyte the solvent should be non polar and vice versa. Some of the commonly used solvents in order of increasing polarity are given below:

Table 3.3: Characteristics of solvents

Polarity	Solvents
Non-polar	petroleum ether
↓	n-hexane
	cyclohexane
	carbon tetrachloride
	trichloroethylene
	benzene
	methylene chloride
	chloroform
	ethyl ether
	ethyl acetate
	acetone
	n-propanol
	ethyl or methyl alcohol
Polar	water

3.1.4.2 Different Extraction Techniques

3.1.4.2.1 Liquid-Liquid Extraction

The most commonly used approach for the extraction of analytes from aqueous samples is liquid-liquid extraction (LLE) (Dean, 1998). The principle of LLE is that the sample is distributed or partitioned between two immiscible solvent in which the analyte and matrix have different solubility or it is based on the low value of the partition coefficient for most organic compounds between water and organic solvents. The main advantage of this approach is the wide availability of pure, solvents and use of low cost apparatus.

3.1.4.2.2 Soxhlet Extraction

Dry materials such as soil, cereal flour and ground plant material can be extracted by this technique. Contact time of solvent and sample matrix varies from 8-24 hours. Although extraction is most efficient, sometimes formation of fine capillary tubes in the sample mass obstructs complete extraction. This technique is not suitable for extraction of heat labile compounds.

3.1.4.2.3 Extraction by Rotating or Shaking with a Solvent/Mixture of Solvents (Tumbling)

Major advantage of this technique of extraction is inclusion of minimum plant co-extractives. This method is often applied to raw fruit or vegetable crops with only a surface residue when pesticide has not been absorbed by plant tissues.

The plant material is cut into approximately 0.5 to 1 cm pieces, placed in extractor with volume of solvent usually equal to or sometimes two times the weight of raw crop. The mixture is shaken or rotated for about an hour to establish intimate contact between the solvent and the crop.

3.1.4.2.4 Extraction by Blending

The material chopped into small pieces is blended with suitable solvent system in a warring blender, preferably explosion-proof, for 2 minutes. The method ensures close contact of solvent and the substrate but suffers from extracting excessive co-extractives along with the toxicant.

3.1.4.2.5 Combination of Tumbling and Blending

The chopped sample is first blended for 1-2 minutes with a minimum of extraction solvent. Then tumbled for about one hour, after addition of remaining solvent/solvents. Formation of stable emulsions during extraction can be avoided by holding excess water with an anhydrous salt such as sodium sulfate or by using co-solvent such as isopropanol which provides solubility for both the extraction solvent and the crop water thereby preventing the formation of a stable emulsion.

3.1.5 Solid Phase Extraction (SPE)

This method of extracting organic pollutants from water, or when ever possible from solid matrices, has been introduced rather recently as an alternative to LLE and Soxhlet extraction. Solid phase extraction (SPE) or sometimes referred as liquid–solid extraction, involves bringing a liquid or gaseous sample in contact with a solid phase called sorbent whereby the analyte is selectively adsorbed on to the surface of the solid phase (Moors et al. 1994). SPE is less labor intensive, produce low background interferences and also significantly reduce the use of organic contaminants. The solid phase is then separated from the solution and other solvents added.

Solid phase extraction separates compounds of interest from impurities in the following 3 ways:

- **Selective Extraction:** In this case, the compounds of interest retained by the packing material and the impurities are eluted out.
- **Selective Washing:** The column is washed with strong solutions to remove impurities but the solution should not be so strong that it carries away the compound.
- **Selective Elution:** In this case the compound of interest is eluted in a solvent but the impurities are retained in the column.

The solid phase/sorbent is usually packed into small tubes or cartridges made of glass or plastic. The thickness of sorbent is usually 0.5 cm. The cartridge resembles a small liquid chromatographic column. For pesticides, commonly used packing materials are silica gel coated with (C_8) or octadecyl (C_{18}) phases, nonporous carbon and florisil. Solvent flow through a single cartridge is typically done using a side arm flask apparatus (Figure 3.3), whereas multiple cartridges can be simultaneously processed by using vacuum manifold.

Method of operation can be divided into five steps. Each step is characterised by the nature and type of solvent used which in turn is dependent upon the characteristics of the sorbent and the sample (Figure 3.4). The five steps are as follows:

(i) wetting the sorbent

(ii) conditioning of the sorbent

(iii) loading of the sample

(iv) rinsing or washing the sorbent to elute extraneous material

(v) elution of the analyte of interest

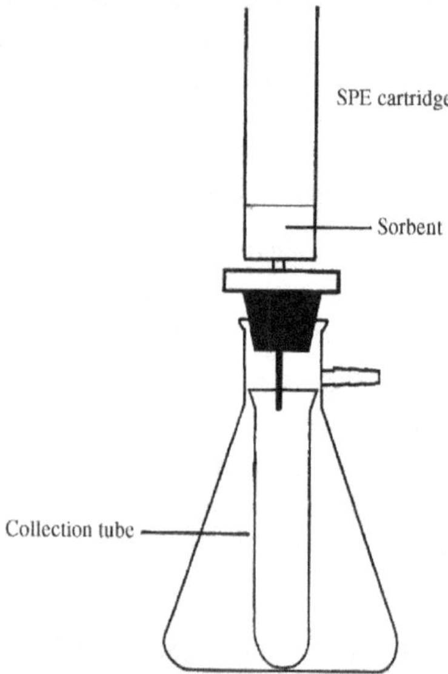

Figure 3.3: Solid phase extraction using a cartridge and a single side arm flask

Wetting the sorbent allows the bonded alkyl chains, which are twisted and collapsed on the surface of the silica, to be solvated so that they spread open to form a bristle. This ensures good contact between the analyte and the sorbent in the adsorption of the analyte step. It is also important that the sorbent remains wet in the following two steps or poor recoveries can result. This is followed by conditioning of the sorbent in which solvent or buffer, similar to the test solution that is to be extracted, is pulled through the sorbent. This is followed by sample loading where the sample is forced through the sorbent material by suction, a vacuum manifold or a plunger. By careful choice of the sorbent, it is anticipated that the analyte of interest will be retained by the sorbent in preference to extraneous material and other related compounds of interest that may be present in the sample. This process is followed by washing with a suitable solvent that allows unwanted extraneous material to be removed without influencing the elution of the analyte of interest. This step is obviously the key to the whole process and is dependent upon the analyte of interest and its interaction with the sorbent material and the choice of solvent to be used. Finally the analyte of interest is eluted from the sorbent using the minimum amount of solvent to affect quantitative release. By careful control of the amount of solvent used in the elution step and the sample volume initially introduced onto the sorbent a pre concentration of the analyte of interest can be affected. Successful SPE obviously requires careful consideration of the nature of the SPE sorbent, the solvent systems to be used and their influence on the analyte of interest. Once developed, the SPE method can then be used to process large quantities of sample with good precision.

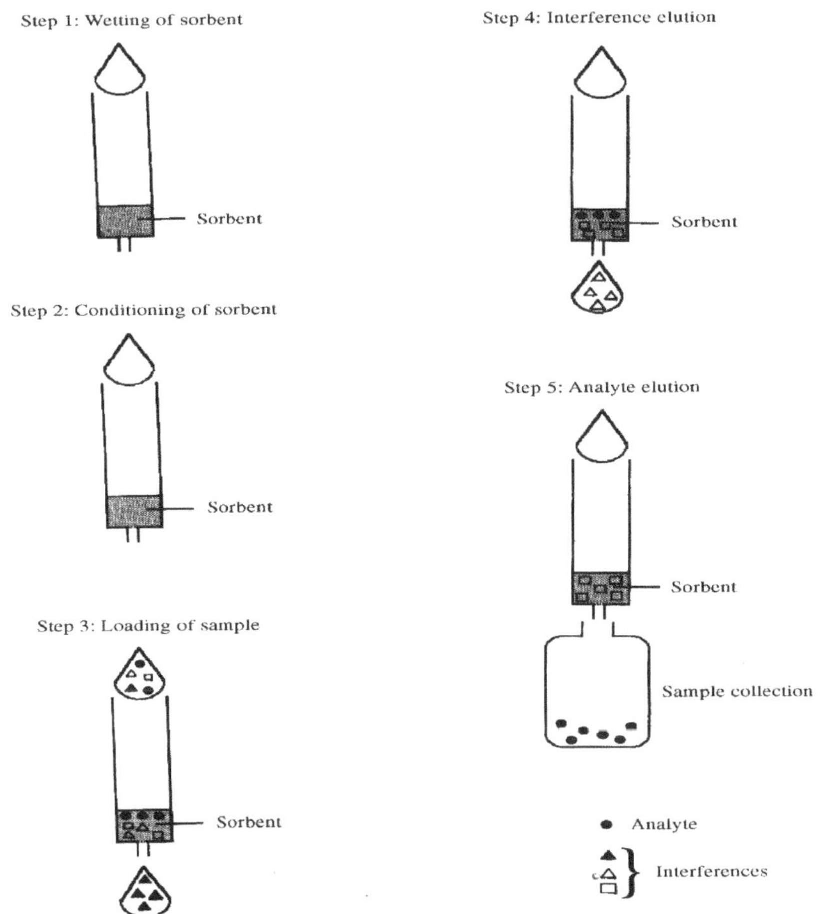

Figure 3.4: Solid phase extraction: Method of operation (Dean, 1998)

Different steps to use the cartridge

3.1.5.1 Selection of Proper SPE tube

Size and capacity of the tube are selected according to the analyte. For less than 1 ml analyte 1 ml tube is selected whereas for 1 ml to 250 ml analyte 3 ml tube is used. The selection of SPE tube depends on:

(i) Sample volume
(ii) Degree of contamination
(iii) Complexity of sample matrix
(iv) Quantity of compounds of interest
(v) Solvent strength of sample matrix

Method of practical operation

3.1.5.2 Conditioning of SPE Tube

SPE tube is conditioned to activate before the sample is extracted. It is ensured that the tube should not dry between conditioning and sample addition in order to avoid the destroying of packing. The choice of conditioning solvent depends on the packing and application. To ensure that the SPE packing does not dry between conditioning and sample addition, the conditioning solvent is allowed to remain at least 1 mm above the top of the tube. Buffer salts are flushed from the tube with water before reintroducing organic solvents.

3.1.5.3 Addition of Sample (Loading)

As the volume of sample can range from micro litre to litres and when the excessive volumes of aqueous solution are extracted, reversed phase packing gradually lose the solvent layer acquired through conditioning. Thus volume of samples is not exceeded beyond 250 ml. To enhance the retention of compound on the packing and elution of unwanted material, pH is adjusted. The sample is passed slowly through the extraction tube using vacuum. Flow rate is not allowed to exceed 5 ml/min.

3.1.5.4 Washing the Cartridge

In order to remove unwanted, weakly retained materials, washing is done with solutions stronger than sample matrix but weaker than needed to remove compounds of interest. Two small aliquots generally elute compounds of interest more efficiently than one larger aliquot. To get the best results normally aliquot is kept in contact with packing for 20 sec. to 1 min.

3.1.5.5 Elution of Compounds of Interest

The packing material is rinsed with a small volume of a solution that removes compounds of interest, but leaves behind any impurities not removed in the wash step. The *eluate* is collected and further processed. Thus by selective extraction, selective washing and selective elution, samples in solution free of interfering matrix components are concentrated for detection.

Most common sorbents are based on silicon particles (irregular shaped particles with diameter between 30 and 60 μm) to which functional group are bonded to surface silanol groups to alter their retentive properties.

Generally SPE can be divided into three classes; normal phase, reversed phase and ion exchange

- **Normal phase sorbent's functional groups:** cyano, amino and diol. It retains polar analytes.
- **Reverse phase sorbent's functional groups:** Octadecyl, octyl and methyl. It retains non polar analytes.
- **Ion exchange sorbent's functional groups:** Have either cationic or anionic functional group and in the ionized form attract compounds of opposite charge. e.g. A cation exchange phase such as benzene sulfonic acid will extract analyte with a positive charge (e.g. phenoxy acid herbicides). In SPE a very short column is fitted with an adsorbent in a fine mesh range (150-400).

Before using this technique following three points are taken into consideration:

Nature of Analyte

Selection of cartridge is done according to the nature of analyte. For non-polar analyte, non polar cartridge is selected. For example, C_8 octa or C_{18} octadecyl phases, non porous graphitized carbon uncoated silica and Florisil material is more appropriate for non polar analytes. In case of polar analytes, sometimes pH has to be adjusted to make the adsorbent's effectiveness. Polar analytes such as carbamates and thiourea pesticides exhibit better recovery with carbon based packing. In general, Envicarb, Oasis having hydrophilic-lipophilic balance (HLB) is used to analyse polar analytes.

Characteristics of Solvents

Here also, phenomenon of like dissolves like is applicable. Solvents are selected according to the characteristics of analytes. The choice of solvent directly influences the retention of the analyte on the sorbent and its subsequent elution, whereas the solvent polarity determines the solvent strength. The solvent strength for normal and reversed phase sorbents is given in Table 3.4.

Table 3.4: Solvent strength for normal and reverse phase

Solvent strength for normal phase sorbents	Solvent	Solvent strength for reversed phase sorben-
Weakest		**Strongest**
	Hexane	
	Iso-octane	
	Carbon tetrachloride	
	Chloroform	
	Methylene Chloride	
	Tetrahydrofuran	
	Diethyl ether	
	Ethyl acetate	
	Acetone	
	Acetonitrile	
	Isopropanol	
	Methanol	
	Polar	
	Water	
Strongest	Acetic acid	**Weakest**

Characteristic of Eluting Solvents

The eluting solvents are selected strong enough to dissolve the analyte to the maximum extent and the interfering materials to the minimum.

3.1.6 Some Considerations Involving Extraction

i) All the solvents should be of analytical grade and should be glass distilled to avoid spurious peaks and low background level. Specially, for electron capture detector (ECD) system, blank sample should be run to check the halogenated impurities. The procedure involved in purification of solvents has been reviewed by Thornburg (1966).

ii) Evaporation of Solvent Solution

It is a very critical step. Evaporation is done by using rotary vacuum flash evaporator or Kuderna Danish evaporator after adding 1-2 drops of mineral oil to arrest the residues. The temperature is usually kept between 40-50° C or according to the boiling point of extracting solvent. During evaporation condensation should be avoided.

iii) Storage of Samples and Extract

Samples having pesticide residues are kept in the deep freeze at a temperature from –25° to –40°C. to avoid degradation of pesticides. Some pesticides such as carbaryl may degrade even at low temperature. For such type of compounds, storing is avoided and analysis is generally carried out on the same day.

iv) Pesticide Recovery or Extraction Efficiency

The degree of residue removal is referred as extraction efficiency and is represented as percent recovery. Extraction efficiency of field incurred residues can only be conveniently evaluated using radio labeled compounds. How many extractions will be sufficient depend on p-valve (fraction of total partitioning in upper phase) Bowman and Beroza (1965).

3.1.7 Clean-up

In order to avoid interferences by co-extractives such as plant pigments, waxes and fats present in the extract, isolation of the toxicant is achieved before taking up estimation of pesticide residues. To achieve the necessary sensitivity, the interfering substances have to be removed from the pesticides with one or more procedures that are commonly referred to as clean-up. Thus, clean-up is a procedure which is used in residue methodology for the isolation of pesticides from interfering extraneous substances. The degree of clean-up required is dependent on the scope of the analysis, the complexity of the sample, and the sensitivity and selectivity of detection methods available for the contaminant sought (Handa et al., 1999, Singh, 2000).

Clean-up usually begins with some form of extraction technique that in the ideal case is capable of completely recovering the analytes of interest while at the same time eliminating the bulk of the sample matrix. Various additional procedures that further fractionate, isolate, purify and concentrate the desired contaminants are then applied as needed for the final determination. While these procedures are generally effective in eliminating undesired sample material, concurrent loss of desired analytes and introduction of interferences from reagents, equipment, and the working environment are potential dangers that must be persistently measured and dealt with as necessary to ensure the validity of results.

Procedures developed for one type of compounds may not be appropriate for another class, or even for all compounds of the same class. Metabolites and degradation products of interest, even though closely related to the parent compounds, sometimes differ significantly in their analytical behaviour. Recovery studies on standard materials carried through the various clean-up steps of a method are necessary to assure reliable determination of previously untested compounds. Sometimes modifications of existing methods or development of new ones may be required as the level of quantitative concern for a chemical contaminant is lowered, the demand for sample clean-up is increased.

Environment samples range in complexity from relatively simple matrices such as air and water to the more complex plant and animal substrates and products derived from them. The extent of clean-up required prior to final determination may be very minimal in some of the simpler cases. More complex samples may require lengthy procedures, often involving the use of a combination of several sophisticated physical and chemical separation techniques. Samples containing large amounts of fatty material can be especially troublesome. Recovery of polar compounds requires the use of polar extraction solvents, which may in turn introduce more undesired co-extractives. A clean-up procedure suitable for the analysis of a particular sample for a compound or class of compounds may not be suitable for those same compounds in another type of sample or for different but closely related compounds in same sample. Hence studies are necessary to:

i) assure reasonably high and reproducible recoveries of standard materials added to the sample matrix

ii) to demonstrate the absence of interferences from sample co-extractives in the final determination are essential when applying clean-up procedures

Developments in detection system with ever-increasing sensitive methodologies and equipments have been dramatic in recent years. Considerable reduction in sample clean-up can be realized, as analytes can often be resolved from undesirable co-extractives either by high resolution chromatographic separation or by detector discrimination. The use of clean-up procedures in excess of what is required for reliable residue analysis is inefficient, and streamlining of the clean-up to the maximum extent allowed by the detection method available is a justifiable goal. However; repetitive analyses of sample extracts containing appreciable amounts of oils, fats, waxes, insoluble particulates, etc. will gradually, if not rapidly, take their toll on most of these detection systems. Basically there are four approaches (Kalra, 1980) to clean-up:

i) Clean-up required when the insecticide is evaluated by direct measurement.

ii) Clean-up required when the insecticide can be measured only after all the interfering substances have been removed.

iii) Clean-up required when the insecticide can be measured after being converted to a suitable derivative.

iv) Clean-up required when methods of compensation can be used.

Thus, the type and extent of clean-up will depend to a great extent on the final method of pesticide analysis. Even though some types of samples can be analysed with little or no clean-up, it may still be advisable to adopt clean-up steps to prevent undue contamination of gas liquid chromatograph (GLC) columns and detectors. The clean-up procedures should be tested to see that they enable detection and determination of the pesticide, being sought, at the desired sensitivity or limit of detection with reasonable recovery (>85%) and with reasonable reduction of background interferences. Every substrate to be considered for insecticide residue study must be individually investigated and quantitatively evaluated as to the performance in the final analytical method. The analyst must choose the clean-up procedure or combination of procedures that will be most practical terms of cost, time and reagent availability (Kalra, 1980).Usually following clean-up techniques are used in pesticide residue analysis.

3.1.7.1 Liquid-Liquid Partitioning

This method is generally adopted for the removal of water soluble impurities. Salting out effect (during partitioning from one solvent to other, dilution is done with saturated aqueous sodium chloride solution) enhance the solubilities of water soluble impurities. Most insecticides have chemical structures foreign to the types of compounds present in the substrate, hence, in fact they also possess some physical and chemical properties unique to the intermediate environment. Such grossly distinctive natures of a pesticide facilitate its separation by favourable partition into one or two suitable immiscible solvents, with the substrate extractives largely favouring the other solvent. Naturally, such manipulations require that the pesticide residues should be preferentially soluble in one solvent, the other solvent should likewise preferentially dissolve the substrate extractives (Cunniff, 1995). Additional requirement is that the solvent in which the insecticide is preferentially dissolved should be low boiling for subsequent ease of removal. The p- values of insecticides in solvent pairs as obtained by Bowman and Beroza (1965) may be of considerable help in making the choice of the solvents.

3.1.7.2 Chromatographic Techniques

It is used to make the extract free from plant pigments. The procedure is based on the adsorption of pigments on acid washed Nuchar (Carbon) and Attaclay or some other adsorbents (Florisil, Celite, Silica Gel, Alumina, Magnesium oxide and Hyflosupercel). The solvent for plant extract is benzene saturated with water. The water modifies adsorbents so that pesticides are not adsorbed strongly. Alternatively mixture of polar and non-polar solvents may be used for elution of pesticides. Kumari et al., (2001, 2002, 2006) used pre washed and activated florisil and activated charcoal mixture for the clean-up of multiresidues of organochlorine and pyrethroids and mixture of silica gel and activated charcoal for the clean-up of multiresidues of organophosphorous and carbamate compounds. Eluting solvents were mixture of ethyl acetate: hexane and acetone: hexane for organochlorine and pyrethroids and organophosphorous and carbamate compounds, from vegetables and fruits extracts, respectively.

3.1.7.3 Adsorption Chromatography

This is the most commonly used chromatographic method for the clean-up of insecticide residues. Cunniff (1995) gave an excellent review of the application of this technique. The most frequently used adsorbents are alumina, charcoal, florisil and silica gel. Columns of two or more adsorbents layered or intimately mixed have also been used occasionally in an effort to combine the best feature of the adsorbents. Such mixed columns pose problems with respect to reproducibility and therefore, a better practice is to use stepwise activation or deactivation of a single adsorbent. Most materials may be activated for a particular purpose by pretreatment with acids or bases (such as alumina) or organic solvents (such as charcoal). Deactivation of alumina is affected by addition of water and different grades of alumina, thus, produced are standardized according to the Brockmann activity standard which is based on the adsorptive capacity or various azo dyes (Greve and Grevenstuk, 1979). Partially deactivated materials are often preferable to active ones. In order to obtain reproducible separation of materials, it is essential to use an adsorbent, well standardized regard to such variables as surface properties, particle size and activity. Universal procedures which can be followed by all workers however require the use of only standardized materials to get reproducible results. The activity of adsorbent can be checked by recovery studies, using a prepared column or by using standard dye material.

Successful separation depends not only upon the adsorbent and on the materials adsorbed but also on the best eluting system. The more strongly bound the solvent the more it will compete with the solutes for the active centers of the adsorbent. Some of the commonly used solvent for this purpose are n-hexane, dichloromethane, chloroform, acetone, methanol, acetonitrile, etc. Ideally, all the solvents should be purified, dried and freshly glass distilled before use.

3.1.7.3.1 Polyethylene Alumina Column

(For removal of fats and waxes) Jones and Riddick (1952) studied the partition of several pesticides between hexane and acetonitrile and found that fats and waxes remained in hexane layer while most pesticides partitioned into acetonitrile layer. Erwin et al., (1955) refined this technique by replacing hexane by paraffin supported on aluminium oxide in a column. The pesticides could then be eluted from the column with 65:35v/v acetonitrile-water mixture while the fats and waxes remained in the column. Unfortunately, some of the paraffin was also eluted from the column and tended to contaminate the sample. Hoskins et al., (1958) substituted polyethylene-coated alumina for paraffin coated alumina. The column is eluted with 65:35v/v acetonitrile-water. The eluate is transferred to a 1 L separatory funnel. Fifty ml benzene/hexane is added to separatory funnel and shaken for 1-2 minute after addition of about 500 ml 4% aqueous sodium sulfate solution. Layers are allowed to separate, the water layer is drawn off and discarded. The benzene/hexane layer is dried with anhydrous sodium sulfate and the extract now purified is ready for further analysis.

3.1.7.3.2 Florisil Column

The extracts of materials containing more than 5 per cent waxes and fats can be cleaned-up by using Florisil column treated with H_2SO_4 provided the toxicant is stable to acidic conditions.

3.1.7.4 Thin Layer Chromatography (TLC)

TLC can be used for clean-up, provided, much fatty material is not present in the extract. Normally, the extract is applied as a streak in this procedure. Standard compounds developed on the same plate are used to indicate the region of the plate on which the pesticide will be found. The residue is removed by scraping and extracted to give a suitable solution for determination (Singh and Kalra, 1987).

3.1.7.5 Channel-Layer Chromatography

Channel layer chromatography is used on the principle of both thin layer and column chromatography. It combines the capacity of column chromatography with the simplicity and speed of thin - layer chromatography. The extract is spotted at the bottom of the plate and developed twice. Plant extractives remain in place or move only slightly while pesticide residues move more rapidly and may be found in solvent front after rechromatography. After drying up, the portion where the insecticide is located is scrapped off from the plate and the insecticide is, then, eluted with a solvent of choice. Faucheux (1968) has successfully used this method to remove co-extractives from several fruits and vegetables.

3.1.7.6 Gel-Permeation Chromatography (GPC)

The most promising technique is gel permeation chromatography (GPC) which is expected to replace the Florisil - based procedures in the future (Thier, 1997). A useful feature of gel chromatography is that the column can be used repeatedly over long periods without any detectable change in the elution volumes and the recoveries. In addition, a miniaturized column of partially deactivated silica gel provides a further cleanup and fractionates the residues according to their polarity, thus yielding extra information for their identification (Anonymous 1991). The separation in gel permeation or gel-filtration or gel filtration chromatography is based primarily on differences in molecular sizes of insecticides and co-extractives. Gel permeation using sephadex LH-20 was shown by Hopper (1981) to be a practical method for the single stage clean-up of insecticide residues from grains. The method was found to be unsatisfactory for the clean-up of cabbage due to co-elution of pigmented material. An improved gel solvent system using Bio-Beads SX-3 gel and tolune -ethyl acetate (1:3v/v) elution solvent was suggested by Johnson et al., (1976). Quantitative recoveries of chlorinated pesticides, organophosphate compounds and polychlorinated biphenyls were obtained with significant savings of both labour and chemicals.

3.1.7.7 Sweep Co-distillation

In sweep co-distillation, the substrate extractive is injected into a glass column packed with glass wool or other suitable material. This is followed by repeated injection of the solvent. Nitrogen carrier gas sweeps the vaporized volatile components through the column, the organic interferences remain on the packing material while pesticides are collected for analysis. A forced volatilization technique was used to obtain vegetable tissue extracts in suitable state of election capture detector (ECD). The sweep co-distillation is a useful method because it eliminates the need for specialized adsorbents and equipment, large volume of costly solvents and laborious cleanup methods (Gunther, 1977).

3.1.7.8 Photodegradation

In this technique, the extract is evaporated to dryness and exposed to high-intensity ultra violet radiations under a germicidal lamp. It is considered that under proper conditions of exposure time and light intensity, many of the non-pesticidal co-extractives will decompose before the pesticide is affected. After exposure, the residue can be redissolved in an appropriate solvent and analysed by gas-liquid chromatography.

3.1.7.9 Coagulation

In this case an aqueous solution of about 0.1 percent ammonium chloride and 0.2 per cent phosphoric acid is added to the filtered solution from the extraction step to precipitate plant waxes, pigments and other interferences which are then removed by filtration. After filtration, the insecticide is partitioned in suitable water immiscible solvent.

3.1.7.10 Chemical Means of Clean-up

If chromatographic purification is not feasible, the extractive interferences can be altered chemically through reactions with acids, bases or oxidizing agents to form products having different solution characteristics from the toxicant.

3.1.7.10.1 Oxidation

Some insecticides extractives may be oxidized then suitably extracted to afford relatively two types of unreactive colourless compounds less likely to interfere with the subsequent measurement of the insecticide. Oxidation can be used to degrade or alter the chemical structure of a compound yielding a fragment or derivative that can be separated from the bulk of plant or animal extractives. For example, conversion of organophosphates to oxy-derivatives for determination by enzymatic inhibition method. The insecticide must be resistant to oxidation.

3.1.7.10.2 Saponification

In case of alkali stable chemicals, saponification with alcoholic caustic is a very effective clean-up technique for high glyceride containing samples. Fats, oils and other high triglyceride containing substrates having pesticide residues can be effectively cleaned by saponification. By refluxing the sample with alcoholic KOH, the triglycerides are converted to soaps and glycerine. The soap solution is extracted with hexane and the pesticide partitioned with the organic phase, while the soaps remain in the aqueous phase. The organic phase is then washed with aqueous alcohol to remove the traces of saponifiable soaps. The greatest disadvantage of saponification is that only alkali resistant pesticides such as aldrin, dieldrin can be cleaned up by this method. Some workers (Veirov and Aharonson, 1978) have used saponification to hydrolyse milk fat and to simultaneously dehydrochlorinate DDT to DDE, which then could be measured rapidly by gas-chromatography.

3.1.7.10.3 Hydrolysis

Hydrolysis with strong acids in columns for plant or animal extracts has been successfully used for bioassay analysis of DDT, BHC (HCH) and lindane (Gunther and Blinn, 1950).

3.1.7.10.4 Reduction

Reducing conditions may be employed to convert a chemical compound derivative or fragment to facilitate its isolation and ultimate measurement.

3.1.8 Physical Means of Clean-up

Physical separation through solvent-solvent partition, steam distillation and freezing or crystallization at low temperatures has also been used for removing excessive amounts of fats and lipids. The use of crystallization or freezing can be used for isolation and concentration of the insecticide. The solid fats could be crystallized out from cooled acetone solutions of acetone-soluble insecticides. Polar and non polar pesticides could be separated from sample lipids, waxes and water by solvent extraction and precipitations at - 80°C. After filtration, the extract becomes suitable for GLC determination.

3.1.9 Estimation Methods of Residues

Several old methods are still in use for the estimation of pesticide residues although quite new reliable and authentic methodology is being in use now-a day for pesticide residue analysis. Some are mentioned below:

3.1.9.1 Bioassay

This method involves the use of sensitive organism (mostly *Drosophila*) whose mortality is directly proportional to the amount of residues. It is good for preliminary information.

3.1.9.2 Spectrophotometry

The active moiety or its derivative is reacted with suitable reagent to make a coloured complex whose intensity is proportional to the amount of residues. Colour intensity is measured spectrophotometrically at a fixed wave length. Although these are old techniques and do not distinguish between parent molecule and its metabolite, nevertheless these methods and being used where costly GC equipment is not available.

3.1.9.3 Chromatography

Thin layer chromatography (TLC), gas chromatography (GC), gas chromatography-mass spectra (GC-MS), high performance liquid chromatography (HPLC), high performance thin layer chromatography (HPTLC) and LC/MS-MS are the latest techniques followed for residue estimations. Both qualitative and quantitative estimations can be made based on perfect resolution through suitable column and versatile/selective detectors.

References

Anonymous, 1991. *Method FSCL Pest-1, Ministry of Agriculture, Fisheries and Food.* Norwich, UK.

Barcelo, D., and Alpenduaruda, M.F., 1996. A review of sample storage and preservation of polar pesticides in water samples. *Chromatographia.* 42: 704.

Bowman, M.C., and Beroza, M., 1965. Extraction p-values of pesticides and related components in six binary solvent systems. *J. Assoc. off. Anal Chem.* 48: 943-952.

Chiron, S., Fernadez-Alba, A., and Barcelo, D., 1993. Comparison of on-line solid phase disk extraction to liquid–liquid extraction for monitoring selected pesticides in environmental waters. *Environ. Sci. Technol.* 27: 2352–2359.

Cunniff, P., 1995. *Official Methods of Analysis of AOAC International,* 16th edn. Arlington VA, l: 1-12.

Dean, J.R., 1998. *Extraction Methods for Environmental Analysis.* John Willey & Sons. Ltd. West Sussex, England.

Erwin,W.R., Schiller, D., and Hoskins,W.M.,1955. Pesicide residue analysis, preassay purification of tissue extracts by wax column. *J. Agric. Fd Chem.* 3(8): 676-679.

FAO, 1986. *Guide to Codex Recommendations Concerning Pesticide Residues.* Part 1: General Notes and Guidelines. Food and Agricultural Organistion, Rome, Italy.

Faucheux, L. J., 1968. Rapid clean-up for carbaryl, using channel layer chromatography. *J. Assoc. off Anal. Chem.* 51: 676-678.

Greve, P.A., and Grevenstuk, W.B.F., 1979. Gas-liquid chromatographic determination of bromide ion in lettuce: Interlaboratory studies. *J. Assoc. off Anal. Chem.* 62: 1155-1159.

Gunther, F.A., and Blinn, R.C., 1950. Persisting of insecticide residues in plant materials. *Ann. Rev. Entomo.*1:167-180.

Gunther, Z.,1977. *Analytical Methods for Pesticides and Plant Growth Regulators.* Vol. VI Academic Press, New York, USA: 763.

Handa, S.K., Agnihotri, N.P., and Kulshrestha, G., 1999. *Pesticide Residues: Significance, Management and Analysis.* Research Periodicals and Book Publishing House, New Delhi.

Hopper, M.L., 1981. Gel permeation systems for removal of fats during analysis of foods for residues of pesticides and herbicides. *J. Assoc. off Anal. Chem.* 64: 720-723.

Hoskins, W. M., Erwin, W. R., Miskus, Raymond, Thornburg, W. W., and Werum L. N., 1958. Pesticide residue analysis, A polyethylene-alumina column for purification of tissue extracts before analysis. *J. Agric. Fd. Chem.* 6(12): 914–916.

Johnson, L.D., Waltz, R.H., Ussary, l. R., and Kaiser, F. E., 1976. Automated gel permeation chrotmatographic clean-up of animal and plant extracts for pesticide residue determination. *J. Assoc. off Anal. Chem.* **59:** 174-187.

Jones, L.R., and Riddick, J.A., 1952. Colorimetric determination of nitroparaffins. *Anal. Chem.* 24: 1533-1536.

Kalra, R.L., 1980. Clean-up procedures for insecticides residue analysis. In: *Residue analysis of insecticide* (ed. Gupta, D.S.), Department of Entomology, HAU, Hisar: 40-56.

Kumari, Beena, Kumar, R., and Kathpal, T.S., 2001.An improved multiresidue procedure for determination of 30 pesticides in vegetables. *Pestic. Res. J.* 13(1): 32-35.

Kumari, Beena, Madan, V.K., and Kathpal T. S., 2006. Monitoring of pesticide residues in fruits. *Environ. Monit. and Assess.* 123: 407-412.

Kumari, Beena, Madan, V.K., Kumar, R. and Kathpal, T.S., 2002. Monitoring of seasonal vegetables for pesticide residues. *Environ. Monit. and Assess.* 74: 263-270.

Marcos, M.P., Chiron, S., Gascon, J., Hammock, B.D., and Barcelo, D., 1995. Validation of two immunoassay methods for environmental monitoring of carbaryl and 1-Naphthol in ground water samples. *Anal. Chim. Acta.* 311: 319.

Moors, M., Massart, D.L., and McDowall, R.D., 1994. Analyte isolation by solid phase extraction (SPE) on silica-bonded phases. Classification and Recommended Practices. *Pure and Appl. Chem.* 66: 277.

Sharma, K. K., 2007. *Pesticide Residue Analysis Method.* Directorate of Information and Publications of Agriculture, New Delhi: 6-10.

Singh, B., 2000. Significance of clean-up in residue analysis. In: *Pesticide residue analysis* (eds. Yadav, P.R., Kathpal, T.S., and Rohilla, H.R.), Department of Entomology, CCSHAU, Hisar: 22-33.

Singh, B., and Kalra, R.L., 1987. Determination of carbofuran residues and metabolites in soil and plant material. In: *4th International Symposium on Analytical Sciences for Environmental Studies,* held at Burdwan (W.B.) from Jan. 23-24, 1987.

Their, H. P., 1997. Harmonization of pesticide residue analytical methods by the Technical Committee, 275 of CEN. *Pestic. Sci.* 50: 151-55.

Thornburg, W.W., 1966.Purification of solvents for pesticide residue analysis. *Residue Rev.* 14:1-11.

Veirov, D., and Aharonson, N., 1978. Simplified fat extraction with sulfuric acid as clean-up procedure for residue determination of chlorinated hydrocarbons in butter. *J. Assoc. off Anal. Chem.*, 61: 253-260.

❑❑❑

Chapter 4
Thin Layer Chromatography (TLC)

Chromatography is an important analytical technique for the separation, isolation and identification of the constituents of a mixture. It is a physical method of separation in which the components to be separated are distributed between two phases; one of these phases consisting a stationary bed of large surface area, the other being a fluid that percolates through or along the stationary bed. All forms of chromatography do have the common features given in this definition. All of them involve the physical movements of the sample across a flat surface (tube like) or through a tube. The stationary phases in all of these possess a large surface area. In all the techniques, the movement of the sample is due to the motion of the fluid/mobile phase. The principle underlying all these methods is the selective retardation of the movement of the components by the stationary phase. These methods result in the production of bands of concentrated components. The basis of naming various chromatographic methods are different:

(a) Paper chromatography is named after the medium used as the solid support.

(b) Adsorption and partition chromatography are named because of the nature of the underlying physical process.

(c) Gas chromatography is named after the state of the fluid phase.

(d) Ascending, descending, radical or two dimensional chromatography is named after the laboratory technique used to carry out the separation.

(e) Column chromatography is named after the container of the stationary phase (Table 4.1).

Table 4.1: Classification of chromatographic methods

Nature of the Distribution Process	Mobile Phase	Stationary Phase	Kind of Chromatography
Partition	Liquid	Liquid	Partition chromatography Paper chromatography
Partition	Gas	Liquid	Gas-liquid chromatography
Adsorption	Liquid	Solid Solid	Adsorption chromatography Thin Layer chromatography
Adsorption	Gas	Solid	Gas- Solid Chromatography

4.1 Thin Layer Chromatography (TLC)

TLC is a chromatographic technique used to separate mixtures. It is a branch of adsorption chromatography. It is based on the difference in adsorption of the components of the given mixture on a given adsorbent. The important factor in TLC is the distribution coefficient of a substance between the stationary phase and the mobile phase. The process is similar to paper chromatography with the advantage of faster runs, better separations, and the choice between different stationary phases. Because of its simplicity and speed, TLC is often used for monitoring chemical reactions and qualitative analysis of reaction products. The stationary phase is the layer of the adsorbent on the glass plate and the mobile phase is the solvent which is made to move through the thin layer of the adsorbent. The components of the mixture or the impurities present in the substance owing to their different absorbability on the thin layer will be distributed stepwise over the latter. The process by which the various components of the mixture move upwards through the thin layer by the continuous flow of the mobile phase is called *elution.* The solvent selected for the purpose is called *eluent.* The component which rises upwards beyond the site of the test sample is called *eluate.* Different compounds in the sample mixture travel at different rates due to the differences in their attraction to the stationary phase, and because of differences in solubility in the solvent.

TLC is one of the most widely used techniques for the separation and identification of pesticides. It has retained favour as an analytical method primarily because of its simplicity, reliability, low cost, and selectivity of detection through the use of various location procedures. Often TLC is only method that can be used for a particular compound. In the analysis of carbamate insecticides for example, it is the method of choice because these compounds are thermally labile. In this chromatographic method a mobile phase moves by capillary action across a uniform thin layer of finely divided stationary phase (adsorbent) bonded on to a plate. When a mixture of pesticides is applied to the plate and developed with the mobile phase, the pesticides move across the plate at different rates depending on their solubility, pK value and capability of hydrogen bonding, and so become separated. The following references will be useful for understanding about TLC completely: Kirchner et al., (1951), Stahl (1969), Michalec (1965),Wise (1967), Lakshminarayana (1980), Sherma (1987), Verlag (1987), Burger and Kaiser (1989), Sherma and Fried (1995), Jaenhen (1997) and Handa et al., (1999).

4.1.1 Its Wide Range of Uses Include

- assaying radiochemical purity of radiopharmaceuticals
- determination of the pigments a plant contains
- detection of pesticides in food
- analysing the dye composition of fibers in forensics
- identifying compounds present in a given substance
- monitoring organic reactions

4.2 Choice of the Adsorbent

For selecting an adsorbent, the polarities of the solvent and that of the sample (to be separated or purified) are to be matched. Another important point is that the adsorbent must not enter into any chemical reaction with the sample or the solvent. The adsorbent should be insoluble and completely inert. The particle size of the adsorbent needs to be preferably in the range 200-300 mesh. Some important adsorbents used in TLC are explained below.

4.2.1 Silica Gel

Silica gel is hydrated silicic acid and is therefore weakly acidic in nature. It is probably the most popular TLC adsorbent. Normally calcium sulphate (gypsum) is added in the silica gel which acts as a binder and the plates are termed *Silica gel G*. The plates have an uniform distribution of particle size, normally about 20 mm in diameter.

4.2.2 Alumina (Aluminium Oxide)

It is slightly basic and contains the carbonates and the bicarbonates of sodium. Presence of these compounds has a marked effect on the adsorptive power of alumina. It can be made neutral by washing with dilute acid, water and finally alcohol. It has fewer applications than silica and needs to be activated to obtain the best separations.

4.2.3 Kieselguhr

This is naturally occurring amorphous silicic acid from the skeletons of diatoms and hence is often referred to as diatomaceous earth. It has less adsorptive properties than silica.

4.2.4 Magnesium Silicate (Florisil)

Florisil is usually used only for specialized separations when other adsorbents have failed.

4.2.5 Cellulose

This uses partition processes for separation, and cellulose plates mimic paper chromatography very well. Although plates made from cellulose fibers are still available, those made from 50 mm spherical particles are preferred. Cellulose plates generally run much more slowly than silica plates of the same thickness.

4.2.6 Ion-Exchange Resins

These TLC plates have cation or anion exchange resins bonded to glass surfaces. The resins are materials such as styrene divinylbenzene copolymers having either quaternary ammonium or sulfonic acid groupings for ion exchange. They are particularly useful for separation of substances of high molecular weight and for amphoteric materials. Strong acids or alkalis are usually used as mobile phases.

4.3 Choice of Solvent

Polarity of the solvent is taken in to consideration while selecting a solvent. Generally the solvent should be of greater polarity then the sample. The choice of the mobile phase, either a single solvent or mixture, depends on the compounds to be separated and the stationary phase to be used. When a stationary phase has been selected, solvents of increasing elution strength can be tried until a particular separation is achieved. Solvents should be cheap and easily obtainable in a pure form. They should be stable in air or when mixed with acids or alkalis, capable of being easily removed from the plate after a chromatographic run, non-toxic, and should not react with the substances to be separated.

4.4 Technique

The basic TLC procedure has largely remained unchanged over the last fifty years.It involves the use of a thin, even sorbent layer, usually about 0.10 to 0.25 mm thick, applied to a firm backing of glass, aluminium or plastic sheet to act as a support. The technique of TLC involves a number of different stages, namely preparing the plate, applying the sample, running the plate, and locating the spots. These stages are described below:

4.4.1 Preparation of Plates

Glass plates of uniform thickness are selected. Removal of all greasy matter from plates is ensured before use. Slurry of the suitable adsorbent in a suitable solvent is prepared by mixing with water and all air bubbles are removed by stirring properly [e.g. in 30 g silica gel G (sufficient for five 20 × 20 cm plates) 60 ml water is added] . Slurry is quickly applied in a 1ayer 250 nm thick to the plates either manually or with commercial slurry spreading equipments, which are of two types:

(i) Moving spreader, stationary plate

(ii) Stationary spreader, moving plates

In both types, the adsorbent is made in to slurry in volatile liquids like water or alcohols and then spread on the supporting plates to give a uniform thickness. The chromatographic properties of a TLC plate may be modified by treating the stationary phase with various agents. They may be added by spraying or dipping the plate in an appropriate solution or by adding the modifier at the time when the plates are being prepared. Simple changes in pH can be effective, e.g. treating silica plates with potassium hydroxide decreases retention and gives more rounded spots for basic drugs. Buffered plates also allow the application of salts or free acids or free bases to the plate. Another example is the use of silver nitrate which can increase the separation of certain compounds, particularly those with carbon-carbon double bonds (this is known as argentation TLC). Other materials that can be useful include ion pairing agents and fluorescent indicators.

4.4.2 Activation of Plates

TLC plates are usually dried in open for 15-20 minutes if binder is present. In the absence of binder, plate is activated at 110 -120^{0}C. Alumina plates are generally heated at 200-250^{0}C for 4 hours. Activated plates are stored either in desiccators or dry place to protect from moisture.

4.4.3 Sample Application

Test sample under examination is dissolved in a suitable solvent to get its clear solution. For this purpose, the solvent used should have the following properties:

1. It should be volatile so that it is removed when the development is started.
2. It should be as non-polar a solvent as possible. Usually, the solvents are graded with hexane as being least polar (Non-polar) and water being most polar (Table 4.2).
3. Before the sample application, a line is drawn with a pencil parallel to, and 2 cm from, the bottom of the plate. The samples are spotted on this line, called the origin, starting 2 cm from the side of the plate and at least 1 cm apart.
4. The sample, normally 1 to 20 mg of material depending on the thickness of the plate, is applied in as small a volume of solvent as possible (usually 1 to 10 μL) and applied by a micro pipette (commercially available products deliver a known volume with an accuracy of 2%), by a capillary tube drawn out to a fine point or by a calibrated micro syringe. Whichever way it is applied, it is vital that the spot is no more than 4 mm in diameter or resolution will be lost.

5. The plate surface must not be cut or gauged by the applicator.
6. The solvent used to apply the spot should be volatile and have low polarity so that the spot does not diffuse too much.
7. The solvent may be applied to the plate in aliquots and dried off naturally or by use of a hot air blower.
8. It is essential that the spot is dry at the end of application, especially if the solution contains water because even a small amount of a polar solvent adsorbed on the plate can drastically alter chromatographic properties.
9. Chromatographic plates are dried after application of spots.

Table 4.2: Commonly used solvents in order of increasing polarity

Non-polar	Petroleum ether
↓	n-hexane
	Cyclohexane
	Carbon tetrachloride
	Trichloroethylene
	Benzene
	Methylene chloride
	Chloroform
	Ethyl ether
	Ethyl acetate
	Acetone
	n-propanol
	Ethyl or methyl alcohol
	Water
	Pyridine
Polar	Organic acids

4.4.4 Development

The capillary action causes the mobile phase to travel through the medium in a process is called development. A saturated developing chamber is normally used consisting of a glass tank which normally has ground edges at the top to make an air-tight seal with the glass lid when coated with soft paraffin. The tank is lined with filter paper on 3 sides and the mobile phase (usually 100 ml for 20 × 20 cm plates) is added. After adding mobile phase, tank is allowed to equilibrate with its vapour for at least 30 min. When the vapour in the tank has equilibrated, chromo plates/TLC plates are immersed in the tank and make it stand in the solvent leaving against the side of the tank. The plates are removed from the tank as soon as the solvent reach the finishing line at the top end of the chromoplates.

4.4.5 Location Procedures

Most of the organic compounds are colourless, and are made visible, preferably by a non-destructive technique. Many compounds are located by examining a plate containing a fluorescent indicator under short wavelength (254 nm) ultraviolet light; absorbing compounds are seen as dark spots on a green background. For long wavelength (350 nm), ultraviolet light is also used, pesticides which naturally fluorescent can be seen. Silver nitrate either built in the silica gel layer or used as a spray has been used widely for the visualization of organochlorine insecticides. Non-destructive tests are performed, by putting the plate in a tank of iodine vapour and brown spots appear. The different adsorbents, developing solvents and chromogenic reagents used for some pesticides used now a days are given in Table 4.3.

4.5 Rf valves

The basic chromatographic measurements of a substance in TLC are the Rf value (Retardation or Retention factor). Once visible, the Rf value of each spot can be determined by dividing the distance traveled by the product by the total distance traveled by the solvent. These values depend on the solvent used and the type of TLC plate, and are physical constants.

$$\text{Rf} = \frac{\text{Distance moved from origin by the compound}}{\text{Distance moved from origin by the solvent front}}$$

- Distance moved from origin by the compound and distances moved from origin by solvent front are recorded.
- From this, the retention factor or Rf value are calculated.

As the distance moved by the solvent front is always more. Hence, Rf value is always less than one. It is important to note that Rf value is constant for each substance but only under identical experimental conditions. The factors on which Rf, value of a substance (component) depends are:

(i) the quantity of the adsorbent material used
(ii) activation grade of two adsorbents employed
(iii) thickness of the adsorbent layer and
(iv) quality of the solvent

4.6 Quantitative Evaluation of Chromatograms

There are two methods for quantitative evaluation of chromatograms. In one method, spot is scratched from the adsorbent and determined using spectrophotometer and gas liquid chromatograph. In second method, spot area is measured in relation to the weight of the material or measurement of density or colour by a photometer.

4.7 General Uses of TLC

- Separation of a great variety of compounds (organic, inorganic, biologicals, polymers, chirals)
- Processing or screening many samples quickly
- Analysis in cases where sample preparation is difficult, undesirable, or impossible
- Determination of analytes in complex matrices
- Post chromatographic derivatization
- Flexible detection, that is, detection in the absence of the mobile phase and with changed parameters, if required

4.8 Common Applications

- Environmental applications from water analysis to plant residues
- Pharmaceutical applications from stability and impurity studies to drug monitoring in biological fluids
- Biomedical compounds (organic acids, lipids, carbohydrates, amino acids, and steroids)
- Food analysis from carcinogens, drug residues, and preservatives to spices and flavors

4.9 Limitations

- Limited to nonvolatile compounds
- Limited to separation numbers (peak capacities) of 10 to 50

4.10 Accuracy

If each step of the technique is automated, precision ranges from 1 to 3 % relative standard deviation (RSD). For non instrumental TLC, relative standard deviations (RSDs) are greater.

4.11 Sensitivity and Detection Limits

- Very much analyte-dependent, with strength of inherent or derivatized visualization/ detection being decisive factors
- Generally sensitivity is in the nanogram range for absorbance and in the picogram range for measurements by fluorescence

Table 4.3: Adsorbents, developing solvents and chromogenic reagents used in thin layer chromatography of some pesticides

Pesticide	Adsorbent	Developing Solvent	Chromogenic Reagent	Colour Developed
Organochlorines				
HCH, DDT, DDE	Al_2O_3 , Silica gel	n-hexane-acetone (8:2v/v)	Silver nitrate-2-phenoxy ethanol	Black spots on off white background
Synthetic pyrethroids				
Fenvalerate, Deltamethrin, Cypermethrin	Silica gel	Hexane-acetone (8:2v/v)	o-dinitrobenzene, p-nitrobenzldehyde	Pink spots
Thiophosphate				
Dimethoate, Methyl-parathion, Coumaphos, Malathion, Parathion, Ethion, Phorate, Disulfoton	Al_2O_3 (Impregnated with DMF)	(i) Methyl cyclohexane (ii) 1,2-Dichloroethane-benzene (iii) Dichloroethane-benzene (1:2v/v)	Tetrabromophenol-phthalein-silver nitrate citric acid 4-(p-Nitrobenzyl)pyridine-tetraethylenepentamine	Blue or purple spots against yellow background Purple blue spots on a white background
Thio-and non thio organophosphates				
Ethion, Phorate, Disulfoton, Methyl-parathion, Parathion, Diazinon	Silica gel	(i) 2,2,4-Trimethyl pentane:acetone:chloroform (70:25:5v/v/v) (ii) 2,6-Dichloroethane-quinione-4-chloroamide	Bromine-Silver nitrate against light brown Thiolo-S and Sulphydryl Thiono-S Thiourea	Dark brown or white spots background Yellow Spots Red Spots Brown Spots

Carbamates				
Carbaryl, Carbofuran	Silica Gel Al_2O_3	(i) H_2O:MeOH (1:1v/v) (ii) H_2O:AcOH:MeOH (5:1:4 v/v) (iii) H_2O: Acetone (3:2v/v) (iv)H_2O:DMF (3:2v/v)	p-niotrobenzene diazonium fluoroborate-KOH Bromine-Fluorescein Rhodamine-B-Ultraviolet	Bluish –reddish Violet Spots, Yellow Spots against orange pink background, Violet spots against orange back ground
Ethyl N-phenylcarbamate, Ethyl N-(nitrophenyl)carbamate, Ethyl-N (2-chloro-4 nitrophenyl) carbamate	Silica gel	(i)Benzene (ii) Benzene: acetone (19:1v/v)	Silver nitrate-ultraviolet	Yellow brown spots against light brown background
Organomercurial Fungicides				
Methyl mercury	Silica Gel Al_2O_3	Hexane: acetone (9:1v/v)	Dithizonates	Red or Yellow Spots
Dithiocarbamate fungicides				
Ziram, Thiram, Maneb, Zineb	Al_2O_3 Silica Gel	(i) Benzene (for dimethyldithio-carbamate) (ii) Benzene :MeOH :AcOH (48:8:4v/v/v) (for ethylene bisdithiocarbamate)	Cuprous chloride NH_4Cl/ NH_2OH. HCl	Bluish Spots on white background
Triazine Herbicides				
Atrazine, Simazine	Al_2O_3	(i) Heptane-benzene-	Sodium azide	Dark Spots

		acetone (10:1:22:5 v/v/v)		
Chlorinated Herbicides				
2,4-Dichlorophenoxyacetic acid	Silica gel	(i) Toluene :acetone (85:15 v/v) (ii) Chloroform-isopropylether (3:2 v/v)	Chlorination Tolidine-KI	Blue Spots on white background
2,4,5 trichloro phenoxy acetic acid, 2,4-dichloro phenoxy butyric acid, 2,3,6-TBA PCP (Pentachlorophenol)2-(2,4,5-trichloro phenoxy propanoic acid)	Polyamide	(i) Acetonitrile saturated hexane (for methyl esters of chlorophenoxy acids) (ii) Cyclohexane: acetic acid (10:1v/v) (for free chlorophenoxy acid)	Silver nitrate-2-phenoxyethanol Chromotropic acid	Black spots on off-white background Violet spot on white back ground

(Adopted from Handa et al., 1999)

References

Burger, K., 1989. Instrumental Thin-Layer Chromatography/Planar Chromatography. In: Proc. *Seventh International Symposium on Planer Chromatography* (Brighton) (Bad Duerkheim, Germany: Institute for Chromatography),UK: 33-44.

Handa, S.K., Agnihotri, N.P., and Kulshrestha, G., 1999. *Pesticide Residues: Significance, Management and Analysis.* Research Periodicals and Book Publishing House, New Delhi: 48-57.

Jaenchen, D.E., 1997.Thin–layer (planar chromatography). In: *Handbook of instrumental techniques for analytical chemistry* (ed. Frank, A. S.) Prentice Hall Inc., New Jersy, US.

Michalec, C., 1965. In: *Stationary phase in paper and thin layer chromatography.* (eds. Macek, K.and Hais, I.M.) Elseveir, Amsterdam, Netherlands.

Sherma, J., 1987. Practice and application of thin-layer chromatography on Whatman KC_{18} reversed phase plates, Whatman Chemical Separations, Clifton, New Jersy, USA.

Sherma, J., and Fried , B., 1995. *Handbook of TLC.* 2nd edn. Marcel Dekker, New York, USA.

Stahl, E., 1969. Thin layer chromatography-A Laboratory Handbook,Springer-Verlag, Berlin, Germany.

Wise, J. J., 1967. *Analytical methods for pesticides, plant growth regulators and food additives.* (ed. Zweig, G.).5: 47.

Kirchner, J.G., Miller, J.M.,and Keller,G.J.,1951. Separation and identification of some terpenes by new chromatographic technique. *Anal.Chem.* 23: 420-425.

Verlag, A.H., 1986. Fundamentals of thin layer chromatography (planar chromatography). Heidelberg, Basel, New York, USA.

Lakshminarayan, V., 1980. Thin layer chromatography in pesticide residue analysis. In: *Residue analysis of insecticide* (ed. Gupta, D.S.), Department of Entomology, HAU, Hisar: 121-129.

Chapter 5

Gas Liquid Chromatography (GC)

Chromatography was first employed by Ramsay (1905) to separate mixtures of gases and vapours. The first scientist to recognize chromatography as an efficient method of separation was the Russian botanist Tswett (1906), who used a simple form of liquid-solid chromatography to separate a number of plant pigments. The colored bands he produced on the adsorbent bed evoked the term chromatography for this type of separation (color writing). Although color has little to do with modern chromatography, the name has persisted and, despite its irrelevance, is still used for all separation techniques that employs the essential requisites for a chromatographic separation, *viz.* a *mobile phase* and a *stationary phase*.

Gas-liquid chromatography (GC) was invented by James and Martin (1952) and is a separation technique in which the mobile phase is a gas (usually helium or nitrogen) and the stationary phase is a liquid. In the original columns used by James and Martin, the liquid stationary phase was adsorbed on the surface of an inert support such as Celite (a diatomateous earth) or calcined Celite (a form of brick dust). The support was usually deactivated before use by acid treatment and subsequent reaction with hexamethyldisilazane. The technique was extensively used for the separation of a wide range of volatile substances. Thus, chromatography has been defined as a separation process that is achieved by distributing the components of a mixture between two phases, a stationary phase and a mobile phase. Those components held preferentially in the stationary phase are retained longer in the system than those that are distributed selectively in the mobile phase. As a consequence, solutes are eluted from the system as local concentrations in the mobile phase in the order of their increasing distribution coefficients with respect to the stationary phase. By separating the sample into individual components, it is easier to identify (qualitate) and measure the amount (quantitate) of the various sample components.

The basis for gas chromatographic separation is the distribution of a sample between two phases. One of these phases is a stationary bed of large surface area, and the other phase is a gas which percolates through the stationary bed. There are numerous chromatographic techniques and corresponding instruments. GC is one of these techniques. It is technique for separating volatile substances by percolating over a stationary phase. Chromatography is probably the most powerful and an versatile technique available to the chemist. It is estimated that GC can analyse 10 - 20% of the known compounds, and of the total pesticides used today, about 80% can be analysed using GC. An analyst can separate gases, and volatile substances by gas chromatography (GC), involatile / nonvolatile chemicals and materials of extremely high molecular weight (including biopolymers) by liquid chromatography (LC) and if necessary very inexpensively by thin layer chromatography (TLC). All the three techniques, GC, LC and TLC have common features that classify them as chromatography systems. The classification of chromatography is given in Table 5. 1.

Table 5.1: The Classification of chromatography

Mobile Phase	Stationary Phase
Gas Gas Chromatography(GC)	Liquid Gas-Liquid Chromatography (GLC)
	Solid Gas-Solid Chromatography (GSC)
Liquid Liquid Chromatography (LC)	Liquid Liquid-Liquid Chromatography (LLC)
	Solid Liquid-Solid Chromatography (LSC)

Several references are useful for understanding the gas liquid chromatogrphy: McNair and Bonelli(1969), Agnihotri (1980),Ravinderanath (1989), Sant (1997), Grob (1995), Dean (1998), Handa et al., (1999), Sharma (2007). Significant technological advances in the area of electronics, computerization and columns have yielded lower and lower detectable limits and more accurate identification of substances through improved resolution because of capillary columns and qualitative analysis techniques.

5.1 What it Does?

Gas chromatographic analysis provides the chemist with a chromatogram and report of the composition of a sample containing a mixture of components. Generally, GC systems are configured and set up for analyzing primarily routine, quality control type samples or for primarily non routine research and development (R&D) analyses. GC systems used for routine analyses can be set up with the latest automation enhancements so that minimum operator assistance is required. One of the major reasons for the widespread use of gas chromatography is the ability to configure the instrument for one's specific needs and budget.

5.1.1 Advantages of GC over the other Analytical Techniques

- Separation can be achieved within minutes.
- Better resolution of closely related compounds.
- Qualitative and quantitative analysis.
- Very sensitive (up to pico and femtogram level).
- Simple to operate and understand the mechanism.
- Data interpretation is very simple.
- Performs dynamic separation and identification of all types of volatile organic compounds and several inorganic permanent gases.
- Performs quantitative and qualitative determination of compounds in mixtures.
- May be destructive or nondestructive, depending on the type of detector used.
- Can determine molecular conformation (structural isomers) and stereochemistry (geometrical isomers) with appropriate detector.
- Can be configured for analysis of specific compounds using a wide choice of options.
- Can be automated for analyses of solid, liquid, and gas samples and may include automated steps in sample preparation.

5.2 Instrument

The basic components of instrument for gas chromatography are illustrated in Figure 5.1. The basic parts of a gas chromatograph are:

(i) Carrier Gas

(ii) Injection port

(iii) Oven/Column

(iv) Detector

(v) Recorder/Database unit

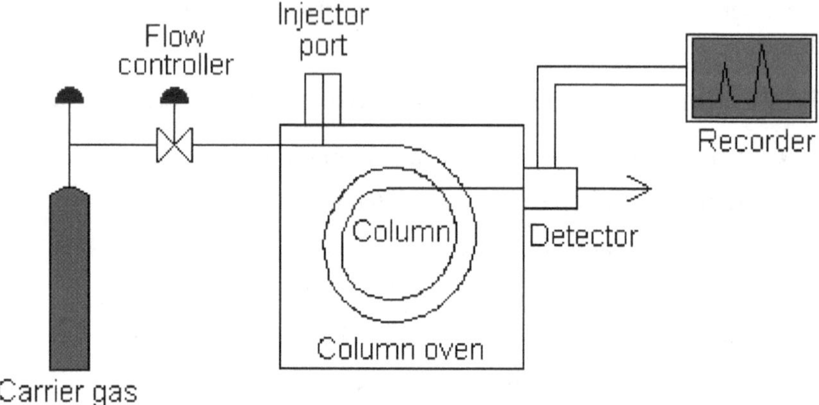

Figure 5.1: Schematic diagram of a gas chromatograph

5.2.1 Carrier Gas

Main function of carrier gas is either to provide a flame or help in burning. Mostly hydrogen and oxygen is used for this purpose. The function of carrier gas is to carry vapours of analyte from injection port to detector through column. The carrier gas used in GC is generally chemically inert. Commonly used gases include nitrogen, helium, argon, and carbon dioxide. The choice of carrier gas is often dependant upon the type of detector which is used. Most commonly used carrier gas is nitrogen. However, hydrogen, helium and argon have also been used.

5.2.1.1 Characteristics of a Carrier Gas

- It should be inert to avoid interaction with sample or stationery phase.
- It should be available in purest form. Carrier gas purity should be better than 99.99% and 99.999% (referred to as five 9s or 99999 grade).
- The gas should be further purified by passing through a series of traps (filters) to remove contaminants (molecular sieve for moisture, charcoal for hydrocarbons and oxygen trap to remove O_2).
- The purifiers must be changed regularly and frequently to prevent saturation; change in colour indicates a requirement for replacement.
- It should be moisture free.
- It should be suitable for the detector being applied in a particular analysis.

5.2.1.2 Carrier Gas Regulators

Normally two stage regulators with stainless steel (ss) diaphragms are used in GC. Although maximum pressure of 80 psi is acceptable but maximum pressure of 200 psi offers flexibility. When carrier gas enters chromatograph, it is usually controlled by one of the three methods use of needle valve, mass flow controller and pressure controller. Most chromatographs available presently have gas flow-rate controller built in the instrument. Flow rate of carrier gas is before gas chromatographic analysis. Small variations in the carrier gas flow rate affect column performance and retention times. For reproducible separations, a constant flow rate is maintained.

5.2.2 Injection Port

Injection port is a device to introduce the sample into the carrier gas stream and the substance to be analysed is injected in solution prepared in organic solvents like hexane or ethyl acetate. Its efficiency is reflected in the overall efficiency of the separation procedure and the accuracy and precision of the qualitative and quantitative results. For optimum column efficiency, the sample should not be too large. Because of ease and convenience, sample is injected with microlitre syringe. Syringe injection method is the most popular method because of the ease and convenience. Special syringes capable of delivering sample volumes in the range of 0.1 - 5000 μl are available. Most commonly used syringes are 1 and 10 μl fixed needle. Samples can be injected either manually or by auto injection. Good reproducibility is achieved with auto injectors than with manual injection. For packed columns, sample size ranges from tenths of a microlitre up to 20 microliters. Capillary columns, on the other hand, need much less sample, typically around 1 μL. Normally for capillary columns, split/splitless injection sytems is used (Figure 5.2).

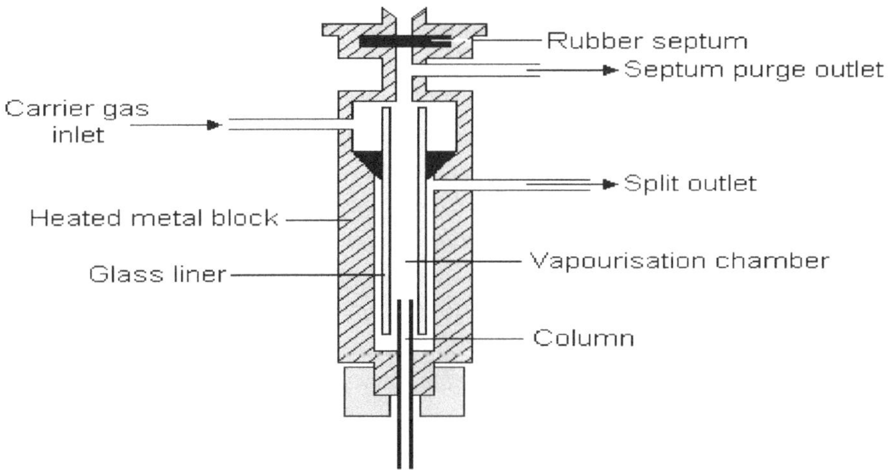

Figure 5.2: Split/splitless injector (Source: http://www.shu.ac.uk)

The injector can be used in one of two modes, split or splitless. A split injection port is designed to allow only a fraction of the injected volume of sample in to the capillary column. The injector contains a heated chamber containing a glass liner into which the sample is injected through the septum. The carrier gas enters the chamber and can leave by three routes (when the injector is in split mode). The sample vapourises to form a mixture of carrier gas, vapourised solvent and vapourised solutes. A proportion of this mixture passes onto the column, but most of the mixture exits through the split outlet. Split ratio is calculated by using the following formula:

$$\text{Split ratio} = \frac{\text{Split vent flow rate} + \text{Column flow rate}}{\text{Column flow rate}}$$

5.2.3 Oven/Column

Chromatographic column is responsible for separation of component in the sample mixture and is called as heart of column. The shape of the column may be straight, bent or coiled. Columns may be made of metal (copper, aluminum and steel), glass and fused silica glass. The length of the column varies from 3-10 feet in case of glass and wide bore whereas it may vary 10-100m in case of capillary column. Efficiency of the column is inversely proportional to diameter of the column. Diameter of the packed column generally ranges from 3-4 mm and it is 320μm for capillary and megabore columns. Film thickness of the stationary phase in capillary column is 0.25μm and 2.65μm is of megabore column. There are two general types of column, *packed* and *capillary* (also known as *open tubular*). Packed columns contain a finely divided, inert, solid support material (commonly based on *diatomaceous earth*) coated with liquid stationary phase. Most packed columns are 1.5 – 10 m in length and have an internal diameter of 2 - 4mm. Packed or porous layer open tubular (PLOT) columns are used for analysis of low molecular weight gases (e.g. CO_2, CH_4, H_2). Capillary columns have an internal diameter of a few tenths of a millimeter. Capillary columns are made from fused silica and synthetic quartz, coated on the outside with polyamide, which make them rugged, flexible and easy to handle. They can be one of two types; *wall-coated open tubular* (WCOT) or *support-coated open tubular* (SCOT). Wall-coated columns consist of a capillary tube whose walls are coated with liquid stationary phase. In support-coated columns, the inner wall of the capillary is lined with a thin layer of support material such as diatomaceous earth, onto which the stationary phase has been adsorbed.

SCOT columns are generally less efficient than WCOT columns. Both types of capillary column are more efficient than packed columns. In 1979, a new type of WCOT column (Figure 5.3) was devised - the *fused silica open tubular* (FSOT) column. Length of capillary columns generally ranges from 10 to 100 m, however, 15 or 30 m columns are commonly used for pesticide residue analysis. These have much thinner walls than the glass capillary columns, and are given strength by the polyimide coating. These columns are flexible and can be wound into coils. They have the advantages of physical strength, flexibility and low reactivity.

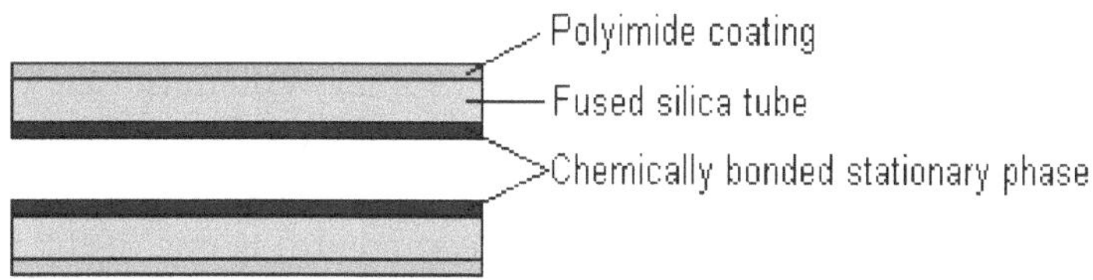

Figure 5.3: Cross section of fused silica open tubular column

(Source: http://www.shu.ac.uk)

5.2.3.1 Solid Support

Purpose of the solid support is to provide large uniform inert surface area for the distribution of liquid phase. It is important to select appropriate stationary phase of columns in optimizing gas chromatographic separation. The stationary phase of column system is chosen after considering polar characteristics of the analytes, their volatility range and column temperature programme. Two main solid supports are Chromosorb- P and Chromosorb-W, the later being more inert and good for polar compounds. Chromosorb is the registered trade mark for solid support material for GC. It is important to select appropriate stationary phase of columns in optimizing gas chromatographic separation. The stationary phase of column system is chosen after considering polar characteristics of the analytes, their volatility range and column temperature programme.

5.2.3.2 Desired Solid Support Properties

- Inert to avoid adsorption
- Thermally stable
- High crushing strength
- Mechanically strong
- Large surface area(1-20m^2/g)
- Smaller particle size to increase column efficiency
- Uniform packing

5.2.4 Liquid Phases

Liquid phases provide differential solubility to components of a substance, which help in their separation. Liquid phases are basically polymeric high melting point silicone greases. About 200 liquid phases are available but only six are most commonly used. Their names are as: SE-30 or OV-101 (dimethyl silicone), OV-17 (50% phenyl methyl silicone), carbowax 20M (polyethylene glycol), DEGS (poly diethylene glycol succinate), Silar-10C (cyanopropyl silicone) and OV-210 (trifluropropyl methyl silicone).Mostly liquid phase is used singly but some times mixed stationary phases are used to achieve selectivity and resolution of the mixture which are difficult to resolve in a single phase.A wide range of stationary phase is available for WCOT capillary columns. One example is a 100 % dimethyl polysiloxane polymer that is chemically bonded onto the interior wall of the column and provides an example of a nonpolar stationary phase. The 5% phenyl polymer composition (middle structure) is one of the most widely used separation phase. DB-5 is first choice of analyst for method development. The stationary phase of DB-5, (nonpolar column) is 5%-phenyl/95% methylpolysiloxane. DB-l has most non polar siloxane stationary phase, surface bonded with 100% dimethyl polysiloxane. DB-5 is similar to DB-l except that 5% of the methyl group is substituted with phenyl group, which is more polar. For medium-polar stationary phase (50%-phenyl-methyl polysiloxane), DB-17 is the best column. Generally the more polar the stationary phase is, better is the separation. Film thickness ranges 0.1 - 5.0 μm, and normally columns with 1.0 or 1.5μm film thickness are used in pesticide determination.

A few common GLC liquid phases used in pesticide residue determination are given in Table 3.2.

Table 5.2: Equivalent commercially available products

Basic structure	**Capillary open tubular**	**Packed**
Polysiloxane, 100% methyl	DB-1, HP-1, HP-101, BP-1	OV-101,OV-1, DC 200
Polysiloxane, 5% phenyl, 95% methyl	DB-5 (ht), HP-5,	OV-3, OV-73,
Polysiloxane, 50% phenyl, 50% methyl	DB-17 (ht), HP-17,	OV-17, OV-11,

(DB, HP and OV are commercial codes for each material are related to their manufacturer) (Adopted from Sharma, 2007)

5.2.4.1 Ideal properties of liquid phases

i) Chemically inert
ii) Thermally stable
iii) Non volatile
iv) Samples must have different distribution coefficients
v) Samples should have reasonable solubility in liquid phase
vi) At operating temperature, liquid phase should have negligible vapour pressure

5.2.4.2 Function of Chromatographic Column

It helps in separation of different constituents of the substance under the influence of a mobile phase.

5.2.4.3 How Separation Takes Place?

By partition co-efficient (k) = $\frac{Cs}{Cg}$

Cs = Concentration in stationery phase

Cg = Concentration in gaseous phase

Narrow ratio means faster separation and vice-versa.

Resolution power of chromatographic column is described in terms of theoretical plates to which it is equivalent to height equivalent to a theoretical plate (HETP), whereas the discrete plate is an artificial concept.

5.2.4.4 How to Measure Theoretical Plates?

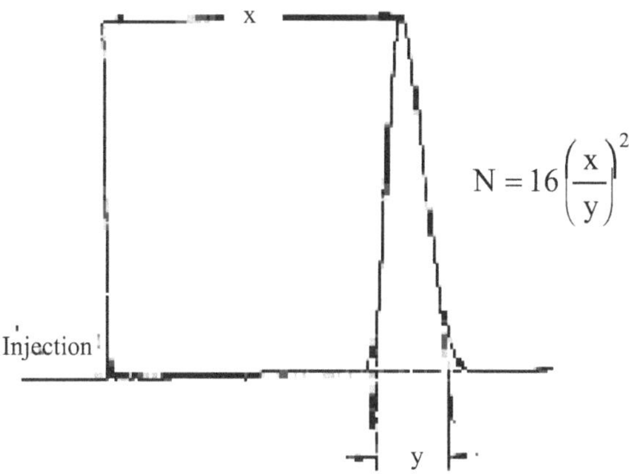

Figure 5.4: Calculation of theoretical plates
(Source: Mcnair and Bonelli,1967)

5.2.4.5 Calculation of Theoretical Plates

Theoretical plates can be measured from the chromatogram. Tangents are drawn to the peak at the points of inflection (about 2/3 of the height) (Figure 5.4). The number of theoretical plates N is given by 16 $(x/y)^2$ where "y" is the length of baseline cut by the two tangents, and "x" is the distance from injection to peak maximum (including the dead-volume). Many factors effect column efficiency and most of these have been evaluated by their effects on N or the height equivalent to a theoretical plate, HETP. This is related to N by:

HEPT = L/N

L is the length of the column (in cm/m). Usually 1000-2000 plates are used for packed columns whereas, 100000 to 1000000 are normal for capillary columns. The more narrow the diameter, greater is the column efficiency or peak sharpness.

5.2.4.6 Resolution

Resolution is a measure of both the column and solvent efficiencies. It accounts for both the narrowness and the separation between maxima (Figure 5.5).

If R=1, the resolution of two equal –area peaks is approximately 98% complete.

If R=1.5, baseline separation (99.7% resolution) is achieved.

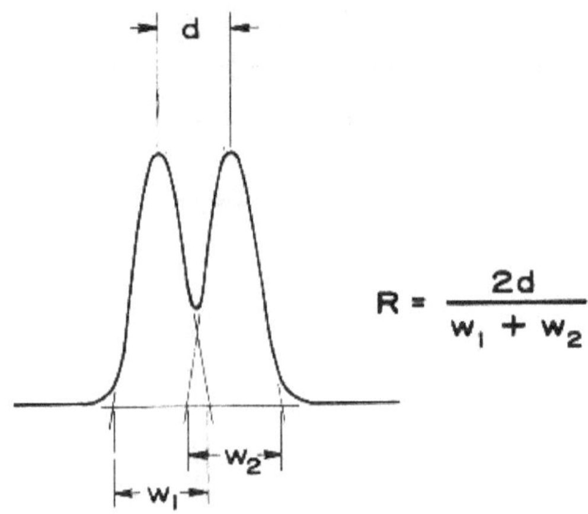

Figure 5.5: Calculation of resolution
(Source: Mcnair and Bonelli, 1967)

5.2.5 Column Temperature

For precise work, column temperature must be controlled to within tenths of a degree. The optimum column temperature is dependant upon the boiling point of the sample. As a rule of thumb, a temperature slightly above the average boiling point of the sample results in an elution time of 2 - 30 minutes. Minimal temperatures give good resolution, but increase elution times. If a sample has a wide boiling range, then temperature programming can be useful. The column temperature is increased (either continuously or in steps) as separation proceeds.

5.2.5.1 Isothermal Conditions

In the simplest GC analysis, the column is maintained at a constant temperature, and this is termed as isothermal analysis. This temperature is selected generally when one or two compounds of similar group and with almost similar boiling points are to be analysed. In this, minimum (say 200°C) temperature is given to oven, 20-25°C higher (225°C) temperature to injector and 20-50°C higher (250°C) to detector. Comparatively, higher temperature of injector and detector avoid the condensation of injected sample. Oven temperature should not exceed the upper temperature limit of the packing material and column should be conditioned for several hours before analysis.

The dependence of GC retention on vapour pressure means that mixtures containing components with a wide range of boiling points cannot be separated satisfactorily in an isothermal run. The more volatile components may well be enough resolved, but higher boiling materials will be eluted with long retention times and very broad peaks. If the column temperature is high enough to give satisfactory peaks for less volatile compounds, the low-boiling constituents will be less well - resolved. The solution is to raise column temperature during a chromatographic run, so that for homologous series, peaks emerge at regular intervals. This is called as temperature programming.

5.2.5.2 Temperature Programming

The retention time of a compound depends not only on the type of stationary phase but also on the column temperature. As the temperature increases, compounds move faster through column and consequently, retention time decreases. Thus, analysis time can be potentially reduced by increasing column temperature. However, sufficient time has to be allowed for compounds to interact with stationary phase to provide the required separation of sample components.

5.2.6 Detectors

Detector is the device that senses the presence of components different from the carrier gas and converts that information to an electrical signal. One's choice of detector includes selectivity and sensitivity. Not all the detectors respond to all components. Selectivity is the ability of the detector to recognize and respond to the components of interest and sensitivity is the concentration level, detected. Sensitivity is defined as the change in the response with the change in detected quantity.

Characteristics of Detector

- High sensitivity
- Low noise level
- Response to all types of compounds
- Inexpensive

Following is the list of the common detectors used for pesticide residue analysis:

- Thermal conductivity detector (TCD)
- Flame ionization detector (FID)

- Electron capture detector (ECD)
- Nitrogen phosphorus detector (NPD)
- Alkali flame ionization detector (AIFD)
- Flame photometric detector (FPD)
- Photo ionization detector (PID)
- Mass selective detector (MSD)

5.2.6.1 Thermal Conductivity Detector (TCD)

It is the oldest type of GC detector used, also known as katharometer or hot-wire detector. Most organic vapors have lower conductivity than H_2, He or N_2 and their presence in one of these carrier gases reduces the amount of heat conducted from the filament. In this, filament temperature increases as analytes present in the carrier gas pass over it, causing the resistance to increase. Although its sensitivity level has dramatically improved through the years, it is still the least sensitive of the detectors. Its popularity has endured because of its ability to respond to any type of constituent that is different from the carrier gas (for example, if helium is the carrier gas, any other component would be detected as long as its concentration is in the nanogram or parts-per-million level, 10^{-9} to 10^{12}).

5.2.6.2 Flame Ionization Detector (FID)

FID is the most commonly used detector in GC. The desirable characteristics of detector which contributed to its status include:

(i) its high and nearly uniform sensitivity (1pg/sec) to most organic compounds

(ii) insensitivity to the common impurities such as carbon dioxide and moisture in the carrier gas

(iii) minimal fluctuations due to changes in the flow, pressure or temperature

Organic components burn in a flame, producing ions that are collected and converted into a current. This is most widely used detector because it offers low detection limits (pico gram or parts per billion). The basic features of FID are shown in Figure 5.6. The FID consists of hydrogen air burner and two electrodes at a potential difference of 100-200 volts. Burner jet is usually negative electrode. The other positive electrode is often a ring, a wire mesh or a cylinder held about a centimeter above the tip of the jet. Carrier gas emerges from the tip of the jet and mix with hydrogen which is burnt in air. Air is introduced into the flame chamber independent of carrier gas and hydrogen. Organic components burn in a flame, producing ions that are collected and converted into a current. This is the most widely used detector because it offers fairly low detection limits picogram or parts-per-billion concentrations) and responds to any type of hydrocarbon component.

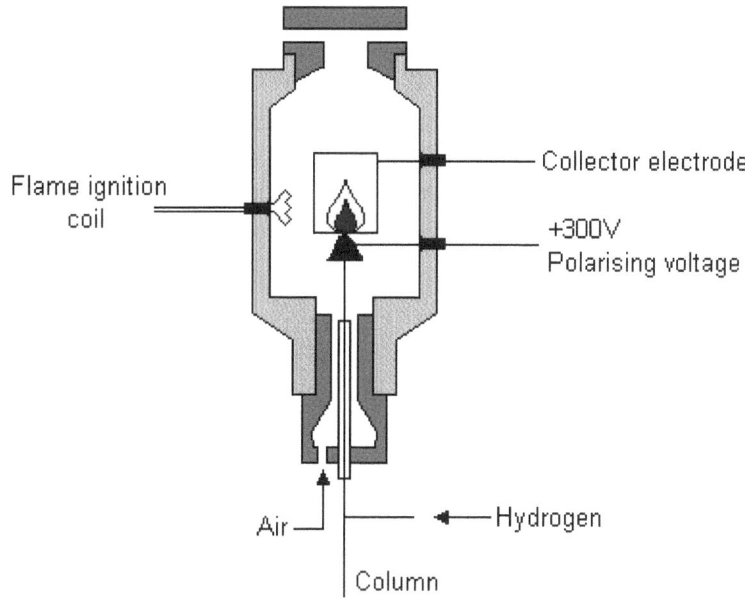

Figure 5.6: Flame Ionization Detector (Source: http://www.shu.ac.uk)

At high temperature of the flame, ionization of hydrogen occurs and current is produced between the electrode. When organic compound containing C-H bonds enters the flame, positively charged ions and electrons are formed.

Organic compound + H_2 + O_2 $\rightarrow$ CO_2 + H_2O + +ve ions + -ve ions + e-

The formation of ions is detected by the passage of current across two electrodes. Increase in current is proportional to the concentration of a compound which is recorded after amplification.

5.2.6.3 Electron Capture Detector (ECD)

The electron-capture detector (ECD) is a highly sensitive detector capable of detecting picogram (pg) amounts of specific types of compounds. The high selectivity of this detector can be a great advantage in certain applications. Its response can vary significantly with temperature, pressure and flow rate. Its principle is to capture of electrons by certain atom of the solute. The most common source of electron is a radio active material which emits low energy beta particles. The detector contains a sealed radioactive source, ^{63}Ni. The radiation ionizes the carrier gas into ion pairs of positive ion and electrons. As the electronegative species pass through the detector, they capture low-energy electrons, causing a decrease in cell current. It is suitable for the very low levels of halogenated pesticides. As its name indicates, its principle is to capture of electrons by certain atom of the solute. The most common source of electron is a radio active material which emits low energy beta particles. The radiation ionizes the carrier gas into ion pairs of positive ion and electrons.

$$N_2 + \beta \rightarrow N_2^{+} + e^{-}$$

This detector contains two electrodes across which a potential is applied to collect and measure the electrons (Figure 5.7). The molecule of the solute capture the electrons, become negatively charged and combine with the positive ion causing decrease in the measurable current which is deflected in the ECD signal.

$$e^{-} + AB \rightarrow AB^{-} \text{ or } e^{-} + AB = A + B^{-}$$

The continued popularity of EC detector (using a ^{63}Ni source) results from its high sensitivity to halogen and certain other moieties as well as its ruggedness and low maintenance needs. Its sensitivity makes it applicable to determination of residues at the ppb and even ppt level. Its wide dynamic response range facilitates its use with automatic data system and its high operating temperature (~400°C) minimizes detector contamination by sample co-extractives and column bleed.

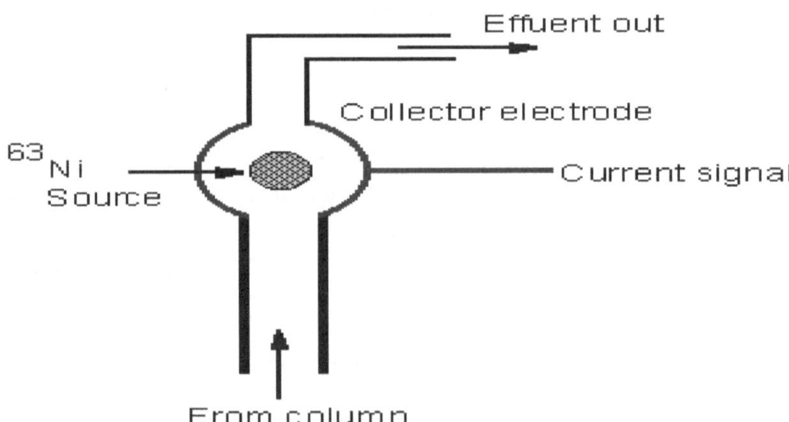

Figure 5.7: Electron Capture Detector

(Source: Sharma, 2007)

5.2.6.4 Nitrogen Phosphorus Detector

This is a detector of choice for analysis of organophosphorous pesticides and pharmaceuticals. Most N-P detectors (NPD) are more responsive to phosphorus than to nitrogen. It is also known as nitrogen-selective detector, because FPD-P is referred for phosphorus residues. Nitrogen and phosphorous compounds produce increased currents in a flame enriched by vaporized alkali metal salt. Response of the N and P detector also provides complementary evidence about element (s) present in the residue; information is often needed for confirmation of identity. In NPD, column effluent impose on to the surface of an electrically heated and polarized alkali source in the presence of air or hydrogen plasma, ionization occurs, and the flow of ions between plasma and an ion collector is amplified and recorded. Detector response to analytes results from increased ionization that occurs when compounds containing nitrogen or phosphorus elute from column. At gas flow rates used for N or P operations, the degree of ionization of compounds containing N or P is > 10,000 times higher than hydrocarbons. All N - P detectors provide electronic heating of alkali source to 600-800^{0}C. This detector can detect as low as 5 -10 pg nitrogen-containing compounds and 1-5 pg phosphorus-containing compounds. Sensitivity of N-P detector is mainly dependent on hydrogen flow and the magnitude of current supplied to the alkali bead. With decrease in hydrogen flow, response to nitrogen increases.

5.2.6.5 Alkali Flame Ionization Detector (AFID) or Thermionic Detector (TID)

Alkali flame ionization detector (AFID) is also known as thermionic detector (TID) or nitrogen phosphorus detector. It is essentially the flame ionization detector modified by introduction of an alkali matter vapour into the system so that its response to nitrogen and phosphorus containing compounds is enhanced by several orders of magnitude. The modification is achieved by fusing an alkali metal salt to negatively charged electrode ring. Such a detector is 600 times more sensitive to phosphorus and 300 times more sensitive to nitrogen containing compounds relative to the unmodified FID .The sensitivity may be further enhanced by putting a second flame detector above the first but separated from it by a salt coated platinum wire mesh. In this design, the first flame burns the organic compounds and vapourizes the alkali metal salt and the vapours are then transferred to the second flame where the ionization occurs.

In the more recent versions the flame is often substituted by electrical heating of the volatile alkali metal activator and hence the name thermionic detector. Several explorations have been offered for the increased sensitivity of the AFID to phosphorus and nitrogen but none very satisfactorily. It appears that the ionization is facilitated by collision of PO, OPO, CN and halogen radicals with the excited alkali metal ions or radicals formed by the application of thermal energy to the metal. Detection limit of AFID is up to pico gram level.

5.2.6.6 Flame Photometric Detector (FPD)

In this detector, sulphur and phosphorus compounds burn in a flame, producing chemiluminescent species that are monitored at selective wave lengths.

5.2.6.7 Photo Ionisation Detector (PID)

This detector contains an ultra violet (UV) lamp. Molecules are ionized by excitation with photons from an UV radiation of sufficient energy (10-12 ev).

$$R + h\nu \rightarrow R^{+} + e^{-}$$

Where R^{+} is the ionized species and hν is a photon having energy equal or more than the ionization potential of the species. The charged particles are then collected in the ionization chamber, producing current and positive current is measured. The PID may be operated in a universal or selective mode by changing the photon energy of the ionization source. Major uses of this detector are for detecting small amounts of aromatics in hydrocarbon mixtures.

5.2.6.8 Mass Selective Detector (MSD)

Molecules are bombarded with electrons, producing ion fragments that pass into the mass filter. The ions are filtered based on their mass/charge ratio. The very popular GC-MS technique yields excellent qualitative identification of components by matching the compounds mass spectrum with spectra included in the libraries that are part of the system.

5.3 Data Acquisition or Identification and Quantification of the Components of Solutes

The last major component of GC system is the data acquisition system. The signal from the detector after amplification is recorded as peak called chromatogram. The area of the peak is representative of the concentration of the component, the larger the concentration, the larger the detector signal and larger will be peak area. The most commonly used devices for generating chromatogram and report are the integrator and the personal computer. Each eluate/solute is identified by its retention time which is defined as the time lapse between the injection of the sample and recording of peak maximum. R_t value is a function of column, temperature, carrier gas flow rate, affinity of the solute for mobile and stationary phase. Under given parameter, the R_t value of the compound remains constant. Quantification of a solute is done by comparing the height or area of each peak of unknown solutes with that of peaks of known standards.

5.3.1 Factors for Ensuring Good Chromatography

- Careful injection of sample into injection port gives good reproducibility.
- Proper rinsing of the syringe to avoid contamination of column and noise peaks.
- Septum should be checked frequently for leakage.
- Temperature of injection port should be at least ten degree higher than column temperature to ensure rapid vaporization of the sample.
- Increase in temperature improve peak shape but tends to bunch rapidly eluting peaks.

- Lowering of temperature leads to increase of retention time and broadening of peak, so the effect on resolution tends to be small, but in case peaks are partially resolved a change of temperature can sometimes have a beneficial effect.

5.3.2 Common Applications

- Separation of compounds in a mixture via a column and detection for qualitative and quantitative identification by any of several types of detectors
- Performs quantitative and qualitative determination of compounds in mixtres
- Used in most types of manufacturing industry for analyses of raw materials, intermediates, or final product
- Used in environmental, forensic, pharmaceutical, and clinical/medical applications

References

Agnihotri, N.P.,1980. Gas chromatography In: *Residue analysis of insecticide* (ed. Gupta, D.S.), Department of Entomology, HAU, Hisar: 130-140.

Dean, J.R., 1998. *Extraction Methods for Environmental Analysis.* John Willey & Sons. Ltd. West Sussex, England.

Handa, S.K., Agnihotri, N.P., and Kulshrestha, G., 1999. *Pesticide Residues: Significance, Management and Analysis.* Research Periodicals and Book Publishing House, New Delhi.

http://www.shu.ac.uk

McNair,H.M., and Bonelli,E.J., 1967. *Basic Chromatography.* Consolidated Printers, Oakland California,USA.

Ravinderanath, B., 1989. *Principles and Practice of Chromatography.* Pub. Ellis Horward Ltd. Chickester, England.

Sant, M.J.V., 1997. Gas chromatography. In: *Hand Book of Instrumental techniques for analytical chemistry.* (ed.Settle, F.A.), Prentice Hall, Inc. New Jersey, USA.

Sharma, K. K., 2007.*Pesticide Residue Analysis Method.* Directorate of Information and Publications of Agriculture, New Delhi.

Chapter 6
High Performance Liquid Chromatography (HPLC)

High performance liquid chromatography (HPLC) was developed during late 1960s and early 1970s. Today it is a widely accepted separation technique for both sample analysis and purification in a variety of areas. HPLC, basically a highly improved form of column chromatography and popularly known as liquid chromatography, like other forms of chromatography, whether paper, thin layer or gas chromatography, needs two essential components, the stationary and the mobile phase. It has the unique ability to separate and quantitate residues of polar, non volatile and heat labile chemical compounds. HPLC uses high pressure pumps up to 400 atmospheres that makes it much faster, short and narrow columns packed with micro particulate phase and a detector. To affect a separation, mixture to be analysed is dissolved in a solvent (ideally this solvent is identical to mobile phase) and then forced to flow through analytical column along with mobile phase. In the column the constituents of mixture get resolved and are detected with a suitable detector by a proper choice of column, mobile phase and detector. HPLC is a technique having high degree of versatility. There are several excellent texts published over the last decade which may be consulted for general information on the instrumentation in liquid chromatography: Hubner (1978), Parris (1984), Ravindranath (1989), Lindsay (1992), Snyder and Kirkland (1996), Brown and Deantonis (1997), Handa et al., (1999) and Sharma (2007).

6.1 The Column and the Solvent

There are two variants in use in HPLC depending on the relative polarity of the solvent and the stationary phase.

6.1.1 Normal Phase HPLC

The stationary phase is polar (hydrophilic).The normal phase packing consists of common porous adsorbents such as silica and alumina. This is essentially just the same as of thin layer chromatography or column chromatography. Although it is described as *normal,* it isn't the most commonly used form of HPLC. The column is filled with a stationary phase of tiny silica particles, and the solvent is non-polar like hexane and iso-octane. A typical column has an internal diameter of 4.6 mm (and may be less than that), and a length of 150 to 250 mm. The least polar analytes elute first, then moderately polar and finally highly polar. Polar compounds in the mixture being passed through the column will stick longer to the polar silica than non-polar compounds will. The non-polar ones will therefore pass more quickly through the column. By decreasing mobile phase polarity, analyte retention can be increased.

6.1.1.1 Limitations of Normal Phase

(i) Irreversible retention

(ii) Long column regeneration

(iii) Water deactivates silica

6.1.2 Reversed Phase (RP) HPLC

The stationary phase is non- polar (hydrophobic). It is most widely used chromatographic mode to separate the neutral molecules in a solution on the basis of their hydrophobicity. In this case, the column size is the same, but the silica is modified to make it non-polar by attaching long hydrocarbon chains to its surface - typically with either 8 or 18 carbon atoms in them. A polar solvent is used - for example, a mixture of water and an alcohol such as methanol. In this case, there will be a strong attraction between the polar solvent and polar molecules in the mixture being passed through the column. There is not as much attraction between the hydrocarbon chains attached to the silica (the stationary phase) and the polar molecules in the solution. Polar molecules in the mixture will therefore spend most of their time moving with the solvent. Non-polar compounds in the mixture will tend to form attractions with the hydrocarbon groups because of van der Waals dispersion forces. They will also be less soluble in the solvent because of the need to break hydrogen bonds as they squeeze in between the water or methanol molecules, for example. They therefore spend less time in solution in the solvent and this will slow them down on their way through the column. That means that now it is the polar molecules that will travel through the column more quickly. Reversed phase HPLC is the most commonly used form of HPLC. Increasing mobile phase polarity increases analyte retention.

6.2 Mechanisms of Separation

Separations by HPLC are achieved using 5 basic operational modes.

1. Liquid solid chromatography
2. Liquid- liquid chromatography
3. Bonded phase chromatography -(i) Ion suppression (ii) Ion pair

4. Ion exchange chromatography- (i) Strong base anion exchange (ii) Strong acid anion exchange
5. Size exclusion chromatography- (i) Gel permeation chromatography (ii) Gel filteration chromatography

The choice mode for a particular application depends on the properties of analyte(s). For pesticide residue determination, bonded phase chromatography is the most widely used mode.

6.2.1 Liquid Solid Chromatography

It is also called as adsorption chromatography. Its stationary phase is solid, silica rich in hydroxyl group. The analyte interact with the stationary phase according to the nature "Like Likes Like". Polar solutes will be retained longest by polar stationary phase and non polar solutes will be retained best by non-polar stationary phases. The basis for separation is the selective adsorption of polar compounds, therefore, the analytes, that are more polar will be attracted more strongly to active silica gel sites. The solvent strength of the mobile phase determines rate of distribution and elution of adsorbed analytes. It is useful for separation of isomers and classes of compounds differing in polarity and number of functional groups.

6.2.2 Liquid -Liquid Chromatography

Liquid liquid chromatography (LLC), also called as partition chromatography involves a solid support usually silica gel or Kieselguhr coated with a film of an organic liquid. In this chromatography, stationary phase is considered to be a neutral liquid and mobile phase is also liquid. Analytes are separated by partitioning between two phases. Components that are more soluble in the stationary liquid phase migrate more slowly and elute later.

6.2.3 Bonded Phase Chromatography

Bonded phase chromatography (BPC) uses a stationary phase that is chemically bonded to silica gel by reaction of silanol groups with a substituted organosilane. Specialized applications of BPC have been developed for ionic compounds, which are highly water-soluble and generally, not well retained on reverse phase (RP) BPC columns. Retention and resolution (separation) can be increased by adding an appropriate pH buffer to suppress ionization (ion suppression chromatography) or by forming a lipophilic ion pair (ion pair chromatography).

6.2.4 Ion Exchange Chromatography

Ion exchange chromatography is used to separate ionic compounds and is based on the principle that opposites attraction. Separation occurs on the basis of the difference in the ionic charge of the sample components. The matrix contains fixed charged groups and the mobile phase counter-ions of opposite charge. Different charged components will have different affinity for the fixed stationary phase charge, thus it is possible to separate mixtures of ionic compounds. The charged nature of the sample components is dependent on the pH of the mobile phase. In this, the stationary phase has an ionically charged surface opposite to sample charge. Mobile phase is an aqueous buffer where pH and polarity are used to control elution. Negatively charged sulphonic acid groups chemically bound to support produce strong acid cation exchange phases. Positively charged quaternary ammonium ions bound to support produce strong base anion exchange (SAX) phases.

6.2.5 Size Exclusion Chromatography (SEC)

In SEC, the stationary phase particles are manufactured with wide range of pore sizes which causes a stationary phase to behave like a molecular sieve. In size exclusion chromatography (SEC), the stationary phase is a material having precisely controlled pore size and the sample is simply screened or filtered according to molecular size as it is moved down the column. As a result of the sieving action, the solutes are separated on the basis of size that big ones come out first. The main difference in SEC and other chromatographic methods is that in SEC, the separation is not achieved by any kind of interaction between the stationary phase and the mobile phase but simply by classifying molecules by size. The analytical columns used for SEC are therefore packed with material having approximately the same particle size as the sample. In SEC, it is of prime importance that the mobile phase must be a good solvent for the sample. Two important subdivisions of SEC are gel permeation chromatography (GPC) and gel filtration chromatography (GFC). GPC uses organic solvents for organic polymers and analytes. It uses aqueous system to separate biopolymers (proteins, nucleic acids).

6.3 Instrumentation

The basic components of HPLC includes:

(i) Solvent reservoir

(ii) Pump

(iii) Injector

(iv) Column and guard column

(v) Detector

(vi) Recorder and integrator or computerized data processor

(vii) Waste collector

In addition to this, HPLC system is usually equipped with on line filters to remove particulate matter. A flow diagram of HPLC is shown in Figure 6.1.

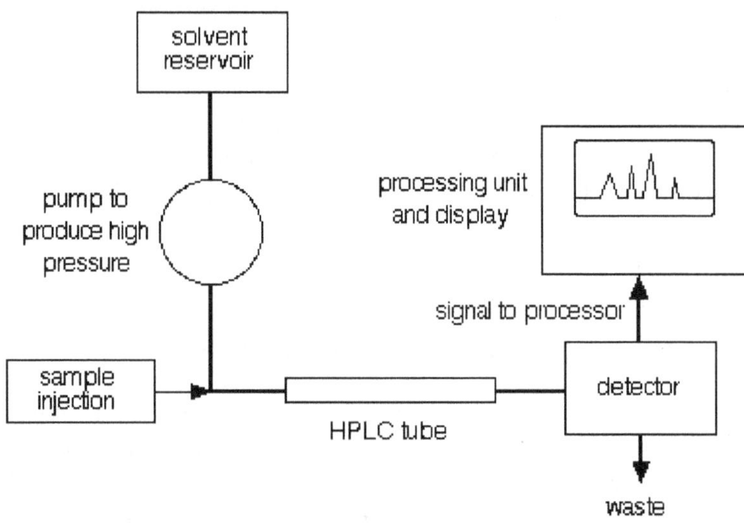

Figure 6.1: A flow diagram for HPLC

6.3.1 Solvent Reservoir

The solvent reservoir may vary from a simple 1 litre Erlenmeyer flask to the very sophisticated solvent storage system and may be made of material from which solvent cannot leach significant impurities. For the purpose of delivering, a solvent should be of uniform composition, free from particulate matter and dissolved gases. The solvent reservoirs may be equipped with magnetic stirrer to keep the composition constant which may be varying due to density differences in the mobile phase components, 2-μm filter to prevent particulate matter from being drawn into the pump, temperature sensor and a heater to maintain a pre-determined temperature or to help degassing and an inert gas (helium) inlet and vacuum connection for degassing. Degassing is particularly necessary in gradient elution where the concentration of the dissolved air in the two solvents differs and when one of the solvent is water (which has a high concentration of dissolved gases relative to other solvents).Removal of oxygen from the system also helps the column life by preventing deterioration of the stationary phase thus helps in increasing column life.

6.3.2 Pumps

The function of the HPLC pumps in the instrument is to deliver mobile phase through column at high pressure with a controlled flow rate. The requirement of a suitable pump at sufficiently high pressure remained in consideration during the early part of development of HPLC. A suitable pump is one such which can deliver precise, reproducible and pulse free flow of solvent through the column at pressure that may range from 500 to 5000 psi. The pumps that can perform to these specifications also need to be resistant to chemical action by the mobile phases and preferably have a low hold up volume for rapid solvent changes as particularly needed in gradient elution. The latest and the most advanced are reciprocating pumps which consist of very small volume chambers (35-400 μl capacity) into which the solvent can be drawn and then pumped into the column by backward and forward movement of a piston or a diaphragm.

6.3.2.1 Gradient Device

A simple one pump HPLC system where mobile phase composition is constant, the method is called isocratic elution. Depending on the complexity of the sample, composition of the mobile phase during analysis to achieve desired separation is to be changed. When this is done, the method is called gradient elution. A gradient employs a change in mobile phase composition over a period of time usually a change in solvent strength with the mobile phase increasing in strength over a set time period. There are essentially two ways to form a gradient. The Beckman way, is to mix the solvents which change the mobile phase composition, under pressure that is after the HPLC pumps. This requires a microprocessor controller to control the delivery of the 2 (binary) or 3 (ternary) pumps employed in gradient elution. Gradient elution is the method of choice when sample components are having widely differing polarities. Change in polarity of the mobile phase during analysis improves separation of the late eluting components and reduces analysis time.

6.3.3 Injectors

The purpose of injection system is to inject sample extract on to the column. The sample is introduced into the chromatography via a typical sample injector, six port injection valve which has two positions i.e. the load and the inject position. Injection valve is the most widely used injection system which introduce sample extract into column without flow interruption. Valves are made with external or internal loops. External loop injectors consist of 6 port injection valves shown in Figure 6.2. The principle of operation of these sampling valves is that the valve is to be set at first in the fill or load position, then sample is introduced into the loop by deep syringe without interrupting the eluent flow. A simple turn of the valve changes the system into the inject position, bringing the sample filled loop into the stream between the pump and the column to prevent the dispersion and broadening of the peaks, the valve is quickly turned to the inject position.

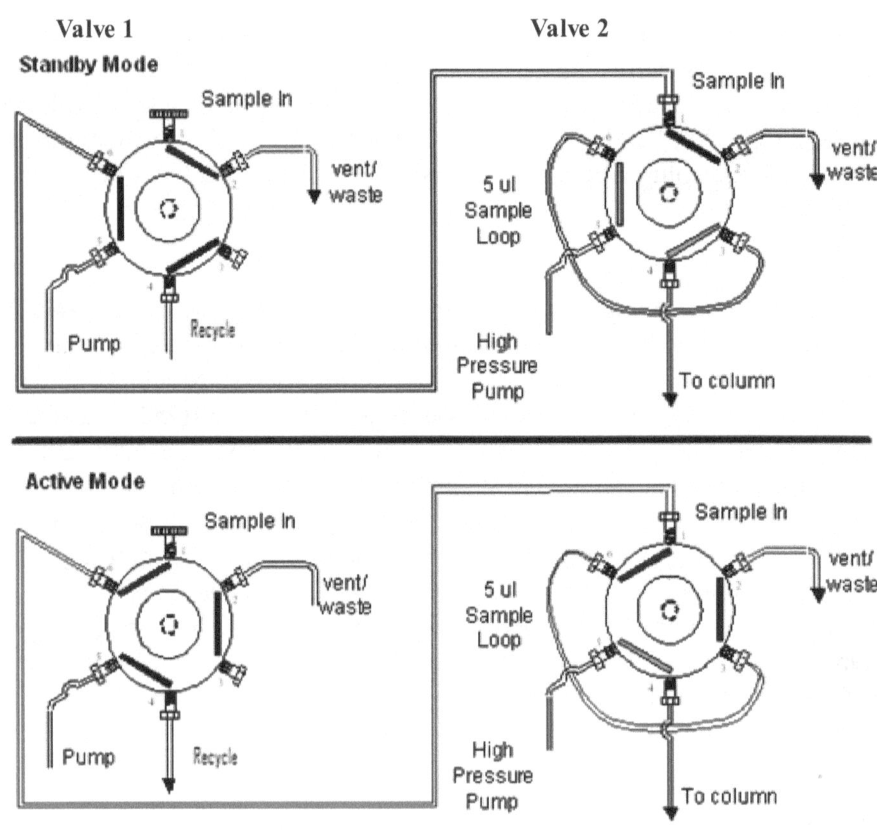

Figure 6.2: Injection valves
(Source: www.leapwiki.com)

Automatic injectors or auto-samplers, are value injectors and very convenient if a large number of samples are to be analysed. They have mechanized syringe to introduce sample in to the loop.

6.3.4 Columns

The column is the heart of the chromatograph. The success of an analysis depends on the selection of column and operating conditions. The column material, the dimensions, the shape, the packing material, the method of packing and the geometry of the end fittings etc. are of decisive importance to the performance of the chromatograph. There are literally hundreds of column packing materials are available for use in HPLC. The packings commonly used in HPLC are listed in Table1. In order of their increasing surface polarity, i.e. from reverse-phase to normal phase packing, followed by the ion exchange packing. The octadecylsilyl (ODS) packings are the most widely used among the reversed phase packings, but better peak symmetry is claimed with octasilyl –bonded (OSB) phase, probably because of their superior wet ability and solute mass transfer characteristics.

Table 6.1: Commonly used HPLC packings

Packing material	Code	Typical application
Octadecylsilyl-bonded silica	ODS	General-pupose RPLC
Octalsilyl-bonded silica	OSB	Solute strongly retained on ODS packings
Phenyl-bonded silica	PBS	Fatty acids,eptides,and solutes with low value on ODS packings
Cynopropyl-bonded silica	CPB	General-pupose RPLC

6.3.4.1 Column Packings and Selection

6.3.4.1.1 Porous Polystyrene

These are spherical highly cross linked resins, capable of high pressure operation and are made by copolymerizing styrene and divinylbenzene as a cross-linking to provide pressure stability (similar to organic size exclusion columns).Degree of cross-linking (4% or 7.5%) influences separation capability. Various ion selective packing available are:

(a) Anion exchange resins normally supplied in chloride form

(b) Cation exchange resins normally supplied in hydrogen form

6.3.4.1.2 Silica Based Ion Exchange Packing Materials

These have following advantages and disadvantages over resin supports.

Advantages

(a) Difference in selectivity, (b) stronger structurally, (c) can use wider variety of solvents without swelling or setting the packing material, (d) can use higher temperature - elevated temperatures (80°C) improve exchange mechanism.

Disadvantages

Narrow pH range, i.e. 2 to 7.5.

Columns may be divided in two groups:

6.3.5 Analytical Column

Pre-packed columns are usually of stainless steel, 3-25-cm long and of 4.6 mm internal diameter (I.D.). Stationary phases used for column packing are uniform, spherical or irregular porous silica gel particles having nominal diameter of 3, 5 or 10 mm. Bonded phases having different functional groups are chemically bonded to surface of silica gel particles. The nature of column packing, together with the nature of the mobile phase largely determines selectivity and efficiency of analysis. The stationary phase in RP mode is hydrophobic and non-polar whereas the mobile phase is relatively polar (usually water with methanol or acetonitrile). Non-polar sample components are strongly retained and polar components are retained less. The HPLC columns are packed with small particles, usually of 3-10 mm, to facilitate pumping of mobile phase at high pressure. Factors important in efficiency of column include narrow particle size, distribution in packing and minimal dead volume in the tubing, fittings, cell and other components of the HPLC system. Recent advances in column technology include use of 3 to 10 cm columns packed with 3 - 5 mm particles. The shorter columns provide faster separation and improved detector sensitivity.

6.3.6 Preparative Column

These columns generally have diameters of 6mm and length from 25-100 cm. Columns are made of stainless steel and are operated at ambient temperatures but elevated temperatures have been used in ion exchange chromatography.

6.3.7 Column Protection

All solvents are filtered through micro porous (0.45 micron) sintered-glass funnels before use. On line filters are also used as extra precautions, to remove the particulate matter from the mobile phase. Guard column of about 3 cm length is inserted between the injection port and the analytical column to prevent clogging of the column. When an acid and a buffer are used as mobile phase, column are to be washed with deionised water and then with methanol. To avoid drying up of column, it is always kept in a proper solvent.

6.3.7.1 Pre-Columns

Pre-columns are positioned in HPLC system prior to sample injector. Purpose of pre-columns is to saturate mobile phase with silica so that silica or bonded silica analytical column does not dissolve during use.

6.3.7.2 Guard Columns

A guard column is fitted between injector and the analytical column to protect the column from damage due to the presence of strongly adsorbed impurities from samples. Guard columns are short (2 - 6 cm) disposable columns containing same packing as analytical column. The use of guard column is quite important when relatively crude sample extracts or biological fluid is to be injected.

6.4 Detectors

The detector converts a change in column effluent into an electrical signal that is recorded by the data station. There are several ways of detecting when a substance has passed through the column. Good detectors exhibit high sensitivity, low noise, a wide linear response range, sensitivity towards pressure, flow and temperature, and response to all types of compounds. Quite a large number of detectors are available for the HPLC.

6.4.1 Refractive Index Detector (RI)

This detector can be called a universal detector for the HPLC as any solute can be detected by this technique as long as there is difference in the refractive index between the solute and the mobile phase. In view of its versatility, it is useful for preliminary screening of unknown samples. However, this detector has certain limitations such as lack of sensitivity, and is not suitable for gradient elution. Its low sensitivity is also an advantage in large scale preparative liquid chromatography. It also requires critical temperature control to operate at highest sensitivity. Refractive index (RI) detectors are non-selective and detect microgram quantities of analytes.

6.4.2 UV-VIS Absorption Detector

The UV-VIS detector is most widely used in the HPLC. It is based on the principle of absorption of UV-visible light as the eluant is passed from the column through a small flow cell held in the radiation beam. This, being a selective detector, is characterized by high sensitivity. The detector is suitable for gradient elution work as it is relatively insensitive to minor changes in temperature and flow rate and also because of the fact that many solvents used in the HPLC do not absorb to any significant extent in the UV-VIS region. Fixed and variable wave length UV/VIS detectors are the most popular detectors. Currently variable wavelength detectors covering the range 190 - 600 nm are widely used.

6.4.3 Fixed Wavelength UV Detector

It is the second most commonly used HPLC detector. Thus, it is less versatile but is much less expensive. It gives much less noise than the variable wavelength detector. Several types of fixed wavelength detectors, differing in the output of the source, are available. These detectors may have

(i) low-pressure mercury lamp source or
(ii) modified mercury lamp with a phosphate and
(iii) medium pressure mercury lamp

6.4.4 Variable Wavelength UV Detectors

Variable wavelength detectors remain the most commonly used detector for HPLC even though there has been a significant increase in the use of fixed wavelength detector and photodiode detector. They usually operate in 190 - 380 nm range with a deuterium lamp or 190 - 900 nm with a tungsten lamp. Other supplementary lamps that have been used are 229 nm cadmium lamp and 214 nm zinc lamp. Variable wavelength detectors provide wide applicability and increased sensitivity. Very pure solvents are required to avoid noise and unstable baselines. An additional advantage of UV-VIS detector is that unknown components can be identified by stopping mobile phase flow and scanning full UV-visible spectrum of the component trapped in the sample cell (stop-flow scanning).

6.4.5 Photodiode Array Detector

Photodiode array (PDA) detectors are now commercially available from major manufacturer of HPLC. The general acceptance and popularity of PDA detector is due to several factors:

(i) it is an information rich detector
(ii) has multivendor availability
(iii) recently available powerful and economical microprocessor for data acquisition and manipulation

The latest in UV /VIS detectors is the photodiode array detector. It functions in an entirely different way than the conventional UV detector. Here polychromatic light is passed through the flow cell and emerging radiation is diffracted by a grating and then falls onto an array of photodiodes, each photodiode receiving a different narrow wavelength band. A microprocessor then scans the array of diodes many times a second and the spectrum so obtained can be displayed or stored. The PDA detector has many advantages over conventional detector e.g. the spectrum of each peak in the chromatogram can be stored and compared with standard spectrum, which facilitates identification of the peaks.

6.4.6 Fluorescence Detector

Fluorescence detector is probably the most sensitive of all the liquid chromatography detectors and is often used in trace analysis of fluorescent compounds. A fluorescence detector can monitor substances those emit light in the visible region after being excited by UV radiation. Fluorescence detector is very specific and its use requires more precautions than any other detector, hence, its use is very limited. Fluorescence is expected in aromatic compounds or compounds of conjugated cyclic structure with a high degree of resonance stability, e.g. polynuclear aromatics, aflatoxins, aromatic amino acids, phenols, quinolines, etc. Polycyclic aromatic compounds with great numbers of delocalized π-electrons have more fluorescence than benzene and benzene derivatives. Electron donating groups e.g.-NH_2,-OH,-F and OCH_3, be liable to enhance the fluorescence, while the electron withdrawing groups such as– Cl, -Br, -I,- NO_2 and –COOH have a tendency to decrease or quench fluorescence.

6.4.7 Photoconductivity Detectors

The photoconductivity detector (PCD) is sensitive and selective for organic halogen, sulphur and nitrogen compounds that form strong, stable ions upon photolysis. The effluent splits as it leaves column, one-half is passed through the reference cell of the conductivity detector and the other half is irradiated with 214 or 254 nm UV light. Suitable analytes become ionized, and the resulting conductance is measured by the detector. Operation of the PCD requires an ion-exchange resin to purify mobile phase and lower background conductivity. Both, reverse phase (RP) and non-aqueous phase (NP) systems have been used with the PCD, but RP are more commonly used for pesticide determination.

6.4.8 Mass Spectrometric Detectors

Mass spectrometric (MS) determination is definitive, and provides information on analyte retention and concentration while simultaneously confirming its identity. Interpretation of mass spectrum permits determination of molecular mass, empirical formula, arrangement of molecular constituents, and ultimately molecular identity.

6.5 Selection of Detector

As many detectors are available, selection of an appropriate detector for a particular application becomes a problem. Most common is variable wavelength UV- visible absorption detector as most compounds do absorb a finite amount of radiation above190 nm or appropriate derivatives may be prepared using either pre or post column derivatization methods.

6.6 Retention Time

The time taken for a particular compound to travel through the column to the detector is known as its retention time. This time is measured from the time at which the sample is injected to the point at which the display shows a maximum peak height for that compound. Different compounds have different retention times. For a particular compound, the retention time will vary depending on:

- the pressure used (because that affects the flow rate of the solvent)
- the nature of the stationary phase (not only what material it is made of, but also particle size)
- the exact composition of the solvent
- the temperature of the column

6.7 Application

It is used for:

(i) Measuring levels of hazardous compounds such as insecticides or pesticides

(ii) Monitoring of environmental samples

(iii) Measuring the levels of active drugs, synthetic byproducts, or degradation products in pharmaceutical industry

(iv) Purifying compounds from mixtures

(v) Separating polymers and determining the molecular weight distribution of the polymers in a mixture

(vi) Measuring levels of certain compounds such as amino acids, proteins and nucleic acids in physiological samples

6.8 Limitations

(i) Resolution can be difficult to attain with complex samples

(ii) Compound identification may be limited unless HPLC is interfaced with mass spectrometry

References

Brown, P., and DeAntonis, K., 1997. High performance liquid chromatography. In: *Handbook of Instrumental Techniques for Analytical Chemistry* (ed. Settle, F.A.),Publishers: Prentice Hall, Inc., New Jersy, USA.pp.147-164.

Handa, S.K., Agnihotri, N.P., and Kulshrestha, G., 1999. *Pesticide Residues: Significance, Management and Analysis*. Research Periodicals and Book Publishing House, New Delhi, India.

Hubner, J.F.K., 1978. *Instrumentation for High Performance Liquid Chromatography.* Elseveir, Amsterdam, Netherlands.

Lindsay, S., 1992. *High performance liquid chromatography*. Wiley, NewYork, USA.

Parris, N.A., 1984. *Instrument Liquid Chromatography*. Elseveir, Amsterdam, Netherlands..

Ravindranath, B.,1989. *Principles and Practice of Chromatography*. Ellis Horwood Ltd., Chichester, West Sussex, England.

Sharma, K. K., 2007. *Pesticide Residue Analysis Method.* Directorate of Information and Publications of Agriculture, New Delhi, India.

Snyder,L.R.,and Kirkland, J.J.,1996.*Introduction to Modern Liquid Chromatography.* 3rd edn.Wiley, New York, USA.

w.w.w.leapwiki.com.

Chapter 7
Supercritical Fluid Chromatography (SFC)

The phenomenon and behavior of supercritical fluid (SCF) has been the subject of research right from 1800's and suggestion of supercritical fluid chromatography (SFC) came in to exixtence in 1958. Supercritical fluid chromatography (SFC) was first developed in 1960, and that time, it was considered as a science fiction chromatography and a revolutionary separation technique. But its reputation has slowly ebbed over the years and it is now moving forward by leaps and bounds as a stable analytical method with many advantages over the existing chromatographic methods. It is an unit operation that exploits the high dissolving power of fluids at temperature and pressurre above their critical values for extracting analytes from sample matrices.The first commercial packed column of SFC was made available in 1981 and the first commercial capillary column SFC instrument was introduced in 1985. Several refrences are useful for understanding the super critical fluid extraction and super critical fluid chromatography: Sandra and David (1980), Smith (1988), Morrisey and Hill,(1989), Christie (1990), Akgerman and Giridhar (1994), McHugh and Krukonis (1994), Taylor (1996),Wenclawiak and Otlervach (2000).

7.1 Supercritical Fluids

Supercritical fluid may be defined from a phase diagram for a pure substance (Figure7.1), in which the regions corresponding to solid, liquid and gaseous state are clear. A substance such as CO_2 can exist in solid, liquid and gaseous phases under various combinations of temperature and pressure. For every substance, there is a temperature above which it can no longer exist as a liquid, no matter how much pressure is applied. Likewise, there is a pressure above which the substance can no longer exist as a gas no matter how high the temperature is raised. Likewise, there is a pressure above which the substance can no longer exist as a gas no matter how high the temperature is raised. These points are called critical temperature and critical pressure, respectively (Smith 1993) and are the defining boundaries on a phase diagram for a pure substance. At this point, the liquid and vapour have the same density and the fluid cannot be liquified by increasing the pressure. Above this point, where no phase change occurs, the substance acts as a supercritical fluid. So SCF can be described as a fluid obtained by heating above the critical temperature and compressing above the critical pressure (Christie,1990). Thus a supercritical fluid is defined as any substance that is above its critical temperature and critical pressure.Critical temperature is therefore the highest temperature at which a liquid can be converted to a liquid by an increase in pressure. Critical pressure is the highest pressure at which a liquid can be converted to a traditional gas by an increase in the liquid temperature. In the critical region, there is only one phase and it posses properties of both gas and liquid. In other words, a gas when compressed isothermally to a pressure more than its critical pressure, exhibits enhanced solvent power in the vicinity of its critical temperature. Such fluids are called supercritical fluids.

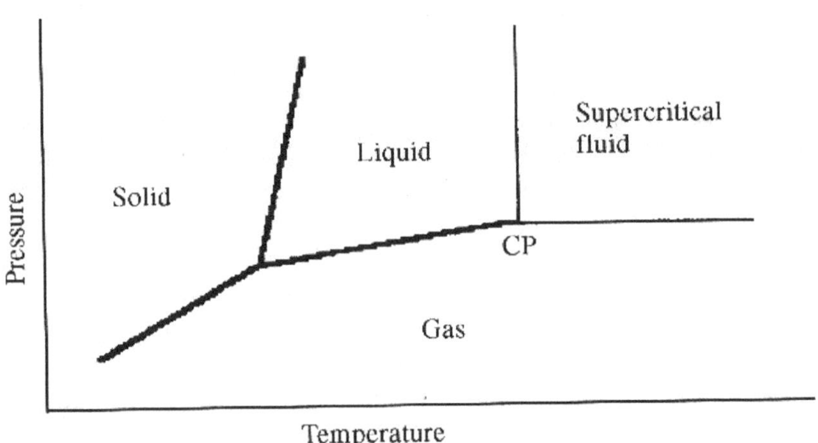

Figure 7.1: Phase diagram for pure substance (Dixon and Jhonston, 1997)

7.2 Important Properties of Supercritical Fluids

The important properties offered by a supercritical fluid for extraction are:-

1. Supercritical fluids have good solvating power. They have high densities (0.2-0.5gm/cm^3) due to which they have a remarkable ability to dissolve large, non-volatile molecules. More density (10^2 to 10^3 times) and low viscosity than gas help in faster dissolution of solute.
2. Supercritical fluids have minimal surface tension.
3. They have high diffusivity and low viscosity. Many SCFs are inexpensive, innocuous, ecofriendly and non-toxic. SCFs have the advantage of higher diffusion constants and lower viscosities relative to liquid solvents. Higher diffusion coefficient means higher analysis speed that increases in the order HPLC, SFC and GC. These advantages are important in both, chromatography and extractions with SCFs.

4. Another important property of SCFs is that, dissolved analytes can be easily recovered by simply allowing the solutions to equilibrate with the atmosphere at low temperatures.

7.3 Supercritical Fluid Extraction

The basic components of an SFE system are: a supply of high purity carbon dioxide; a supply of high purity organic modifier; two pumps; an oven for the extraction vessel; a pressure outlet or restrictor; and a suitable collection vessel for quantitative recovery of extracted analytes (Dean, 1998).The most common modifier used is methanol. The pressurized gas in the cylinder is pumped through the SFE system using a syringe pump. Schematic diagram of a supercritical fluid extraction unit is presented in Figure 7.2. The carbon dioxide and modifier are then subsequently mixed using a T-piece. The combination of two pumps allows a high degree of control in terms of modifier addition and flexibility of choice. The critical temperature for the solvent system is established by external heating. This is done via an oven in which is located the sample cell or vessel. The ideal temperature range of the oven is up to 100°C.To increase the extraction efficiency; samples are cut in small pieces in order to expose more surface area to the extraction solvent. Well chopped or powdered material is loaded/charged into a extraction cell, pressurized, and brought to a temperature of 50 to 75°C. CO_2 is used at desired pressure and temperature 138-345 bar and 40-50°C respectively and solute SC-CO_2 leave the extractor (CO_2 becomes supercritical at 31°C and 72 atmosphere).Once the extraction is complete, the analytes are trapped, recovered, and analysed. Solutes are mostly in ppt levels. Flow of CO_2 is maintained 3-40 kg/h at 20°C.

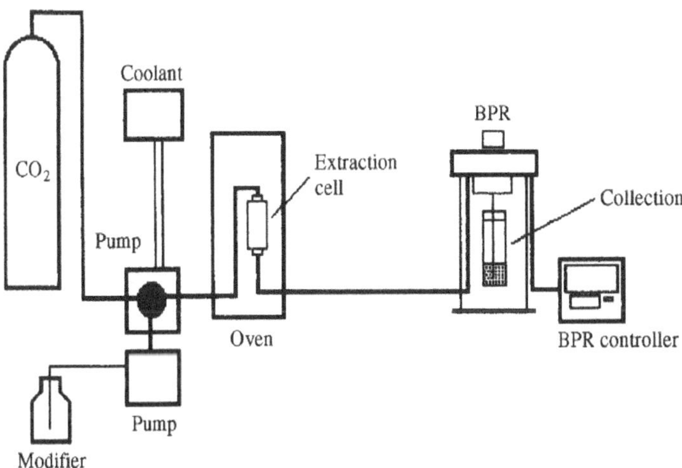

Figure 7.2: Schematic diagram for supercritical fluid extraction system (Dean, 1998)

7.4 Supercritical Fluid Chromatography (SFC)

Chromatography is an analytical technique used for the separation of complex chemical mixtures into individual components. SFC is a relatively recent chromatographic technique, having been commercially available since only about 1982. SFC is a unit operation that exploits the high dissolving power of fluids at temperature and pressure above their critical values for extracting analytes from ample matrices. In SFC, the sample is carried through a separating column by a supercritical fluid where the mixture is divided into unique bands based on the amount of interaction between the individual analytes and the stationary phase in the column. As these bands leave the column, their identities and quantities are determined by a detector. SFC is a hybrid of gas and liquid chromatography because when the mobile phase is below its critical temperature and above its critical pressure, it acts as a liquid, so the technique is liquid chromatography (LC) and when the mobile phase is above its critical temperature and below its critical pressure, it acts as a gas so the technique is gas chromatography (GC) (Raymond Scott,1994). In SFC, the mobile phase is initially pumped as a liquid and is brought into the supercritical region by heating it above its supercritical temperature before it enters the analytical column. It passes through an injection valve where the sample is introduced into the super critical stream and then into the analytical column. It is maintained supercritical as it passes through the column and into the detector by a pressure restrictor placed either after the detector or at the end of the column.

7.4.1 Principles of Operation

Supercritical fluid chromatography (SFC) uses a mobile phase which consists of a highly compressed gas at above its critical temperature and pressure. The physical properties of supercritical fluids are between those of gases and liquid, resulting in good transport of analytes through the chromatographic column. Some non polar compounds like CO_2, NH_3, H_2O, propane and methanol act as supercritical fluids. The two supercritical fluids of particular interest are, carbon dioxide and water.

7.4.1.1 Carbon dioxide

There are a number of possible fluids which may be used in SFC as the mobile phase. However, based on its low cost, low interference with chromatographic detectors, and good physical properties like nontoxic, nonflammable and ecofriendly solvent with low critical temperature of 304K and moderate critical pressure of 73bar, carbon dioxide is the standard. It is miscible with variety of organic solvents and is readily recovered after processing. As it's a small and linear molecule, it diffuses faster than conventional liquid solvents. CO_2 is most widely used as SCF because of its no harmful residues due to colourless, tasteless, invert non-inflammable characteristic non-toxic nature and property facilitate extraction at low temperature in non oxidizing environment. The main disadvantage of carbon dioxide is its inability to elute very polar or ionic compounds.

7.4.1.2 Water

It has a critical temperature of 647K and critical pressure of 220 bar due to its high polarity.

7.5 Basic Parameters of Supercritical Fluid Chromatography

There are four basic parameters of supercritical fluid chromatography (Handa et al.,1999).

1. The first of these is the miscibility pressure which is the pressure at which the solute starts to dissolve in the supercritical fluid. This parameter is known as "threshold pressure" and helps to know the mixing between the dissolved solute and the solvent gas.
2. The other parameter is the pressure at which the solute attains its maximum solubility in the compressed fluid. When the solubility parameter of the extracting fluid (gas) is equivalent to that of the solute, maximum solubility should be attained.
3. The third parameter is the pressure region between the miscibility and solubility maximum pressure, in the fractionation. Pressure range in which a solute's solubility will range between zero and its maximum value in the supercritical gas.
4. The melting point of the solute is a particular parameter in SFE because most solutes are dissolved to a greater extent in the supercritical fluid medium when in their liquid state.

7.6 Advantages of Supercritical Fluid Chromatography

1. Carbon dioxide which is typically used as the supercritical fluid for extraction is non-toxic, leaves no solvent residues, has no disposal costs and is substantially less expensive than liquid solvents.
2. There are no environmental risks and no fire hazards because carbon dioxide is used as solvent.
3. With super critical fluid, extraction can be performed in 5-30 minutes because of low viscosity of SCF.
4. It eliminates vacuum distillation or evaporation steps associated with conventional technique.
5. SFC increases accuracy and reproducibility of extraction.
6. This allows fast analysis of thermo labile compounds.

In the last few years, the analytical application of supercritical fluids has expanded rapidly, mirroring a similar surge of interest in supercritical fluid extraction. Supercritical extraction utilizes the properties of a compound at temperatures and pressures above its critical point. More density (10^2 to 10^3 times) and low viscosity than gas help in faster dissolution of solute. On the basis of current data solubility trends in SCF is as under:

- Polar solutes are most soluble in polar super critical fluids.
- Maximum increase in solubility with pressure usually occur near critical density.
- Under constant density, solubility increases with temperature.

SCF is a powerful separation tool for extracting the bound residues, thermally labile compounds and conjugate food.

Brady et al., (1987) used CO_2 to extract DDT, PCBs and toxaphene from soil. Hopper and King (1991) have applied supercritical fluid carbon dioxide extraction technique for extraction of chlorinated pesticides from foods using palletized diatomaceous earth. Supercritical fluid chromatographic research has also demonstrated that a variety of pesticides can be separated using SFC-CO_2 from soils (Engelhardt and Grass; 1988; and Lopez-Avila et al., 1990) and plant tissues (Hawthorne et al.,1988. Michelangalo et al., (1998) used SFC with CO_2 for the analysis of N-methyl O-aryl carbamate in fruits and vegetables. Supercritical fluid extraction and chromatography has been used for the analysis of pesticide residues in canned foods, fruits and vegetables wherein pyrethroids, herbicides, fungicides and carbamates have been tested (El-Saeid,2003)

7.7 Application

As agricultural and environmental compounds and food products are polar and non volatile, so GC can not be applied to these samples without derivatizing the analytes. HPLC has been a very popular technique for the analysis of polar, nonvolatile and high-molecular weight compounds. SFC offers an attractive alternative for the separation of volatile, nonvolatile and thermally labile material. The greater diffusivity and lower viscosity afforded by SFC relative to a liquid yield faster, more efficient separations and liquid like densities enable man thermally inaccessible components to be solubilized. Its specific advantage over HPLC is greater resolution per unit time and less time. Supercritical fluid extraction chromatography can be used for analysis of:

(i) pesticides in foods, tissue and soils
(ii) pharmaceuticals and drugs
(iii) polymers and polymer additives
(iv) triglycerides, lipids and phospholipids
(v) surfactants
(vi) hydrocarbons and petroleum products.

References

Akgerman, A., and Giridhar, M.,1994. Fundamentals of solid extraction by supercritical fluids, In: *Supercritical Fluids- Fundamentals for applications.* (eds., Sengers J. M. H. and Kiran E.), Klüwer Academic Publishers. 669-696.

Brady, B.O., Kao, C.P.C., Dooley, K. M., Knopf, F.C. and Gambrell, R.P.,1987. Supercritical extraction of toxic organics from soils. *Ind. Eng. Chem. Res.,* 26: 261-268.

Christie, W. W.,1990. Supercritical fluid chromatography and lipids. *Lipid Technol.* 2:107-109.

Dean, J.R., 1998. *Extraction Methods for Environmental Analysis.* John Willey & Sons. Ltd. West Sussex, England.

Dixon, D. J. and Johnston, K.P., 1997. Supercritical fluids. In *Encyclopedia of Separation Technology*. (Ruthven, D.M., ed.), John Wiley, 1544-1569.

El-Saeid, M.H., 2003.Pesticide residues in canned foods, fruits and vegetables: the application of supercritical fluid extraction and chromatographic techniques in the analysis. *Scienti. World J.* 11(3): 1314-1326.

Engelhardt, H., and Grass, A., 1988. *J. High Resdut. Chromatog and Chromatog. Commun.* 11 : 726.

Handa, S. K., Agnihotri, N.P. and Kulshrestha, G., 1999. *Pesticide Residues: Significance, Management and Analysis.* Research periodicals and book Publishing House, New Delhi.

Hawthorne, S.B., Krieger, M.S., and Miller, D.J., 1988. Analysis of flavor and fragrance compounds using supercritical fluid extraction coupled with gas chromatography. *Anal. Chem.* 60(5): 472-477.

Hopper, M.L., King , J.W., 1991. Enhanced supercritical fluid carbon dioxide extraction of pesticides from foods using pelletized diatomaceous earth. *J Assoc. off Chem.* 74: 661-666.

Jentoft, R. E., and Gouw, T. H., 1976.Analysis of polynuclear aromatic hydrocarbons in automobile exhaust by supercritical fluid chromatography. *Anal. Chem.* 48(14): 2195-2200.

Lopez-Avila V, Dodhiwala NS, Beckert WF. 1990. Supercritical fluid extraction and its application to environmental analysis. *J Chromatogr. Sci.*28: 468-476.

McHugh, M., and Krukonis, V., 1994. *Supercritical Fluid Extraction.*2ndedn. Boston: Butterworths, USA.

Michelangalo, A., Ellen, S., Friedrich, M., D., Wolfgang, S., 1998. Analysis of several N-methyl-O-aryl carbamates in fruits and vegetables after super critical fluid extraction. Abst. In: Book of Abstracts. *'9th International Congress of Pesticide Chemistry' The Food and Environment Challenge'* held at Queen Elizabeth II Conference Centre Westminster, London, UK, from 2-7 August,1998.Vol. 2p.7A-044.

Morrisey, M.A., and Hill, H.H.Jr.,1989.Selective determination of underivatized 2,4-dichlorophenoxy acetic acids in soil by supercritical fluid chromatography with ion mobility determination. *J. Chromatogr. Sci.* 27(9): 529-533.

Raymond Scott, P. W., 1994. Liquid Chromatography for the Analyst. Marcel Dekkar: 7.

Sandra P., David, F., 1980.Supercritical fluid chromatography. Smith R. M., Ed., Royal Society of Chemistry, London :137-158.

Smith R.M.,1988. *Supercritical Fluid Chromatography.* The Royal Society of Chemistry, London: 5.

Smith, R.M., 1993.Nomenclature for supercritical chromatography and extraction, IUPAC Recommendations. *Pure and Appl. Chem.* 65(11): 2397-2403.

Taylor, L., 1996. *Supercritical Fluid Extraction.* Wiley,New York: USA.

Wenclawiak, B. and Otlervach, A., 2000.Carbon-based quantitaion of pyrethrins by supercritical fluid chromatography. *J . Biochem. Biophys . Methods.* 43 (1-3): 197-207.

❑❑❑

Chapter 8

High Performance Thin Layer Chromatography (HPTLC)

In the present era, residue analyst is exposed to array of analytical choice and as per International regualtions, presence of pesticide residues in environmental components and food commodities are becoming more stringent.Generally, multiresidue methods are based on GC equipped with selective detection either electron capture detector, nitrogen phosphorous detector, flame ionization detector etc. One most widely applied technique in pesticide residue analysis is HPLC, equipped with different detectors. Another up-date method for confirmation is gas chromatography-mass spectrometry (GC-MS) and HPLC-MS applying particle beam or thermo spray ionisation technique. Recently one more analytical technique which has gained ground in the field of pesticide residue analysis, is high performance thin layer chromatography (HPTLC).The criteria of good analytical technique like simple, rapid, flexible, versatile, inexpensive, easily documented and validated has been the inherent advantage of HPTLC. Strictly speaking, HPTLC is not different in principle or application from conventional TLC, but due to improvements in materials, its effectiveness has also enhanced.The technique has a potential to operate upto femtogram level. HPTLC is a sophisticated and automated form of TLC. It is a technique which can generate about 5000 or more theoretical plates. The high performance in TLC is achieved by optimized plate coating materials, by improved sample application techniques, and by developing new methods for the development of mobile phase. Thus HPTLC is an instrumental approach at different stages of TLC. A flow chart for HPLC principle is shown in Figure 8.1. The principle of separation is based on adsorption and this leads to easy separation of isomers, including chiral compounds. Main difference between HPTLC and TLC is given in Table 8.1

8.1 Main Features of HPTLC

1. Simultaneous processing of sample and standard - better analytical precision and accuracy less need for internal standard
2. Several analysts work simultaneously

3. Less analysis time and less cost per analysis
4. Low maintenance cost
5. Simple sample preparation - handle samples of divergent nature
6. No prior treatment for solvents like filtration and degassing
7. Less mobile phase consumption per sample
8. No interference from previous analysis - fresh stationary and mobile phases for each analysis - no contamination
9. Visual detection possible - open system
10. Non UV absorbing compounds detected by post-chromatographic derivatization

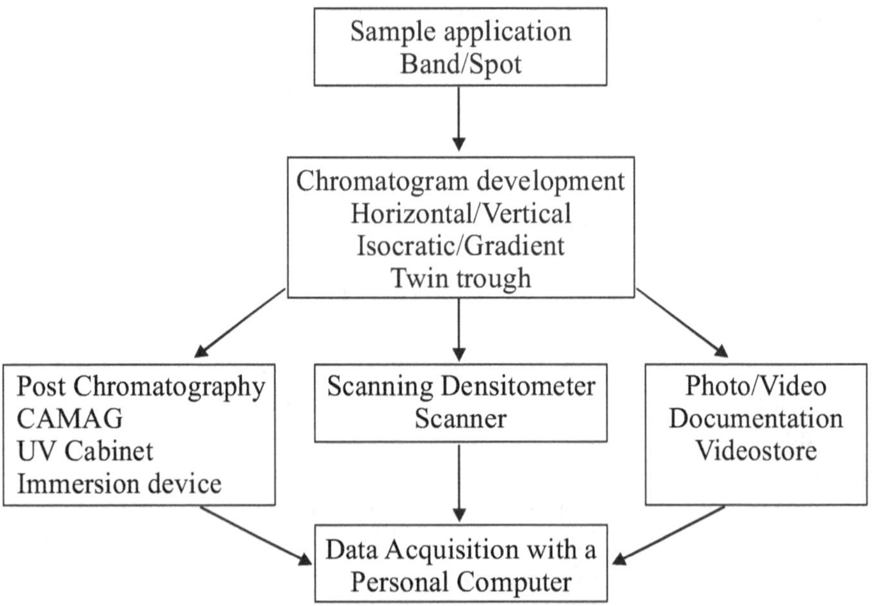

Figure 8.1: Principle of HPTLC

(Source: Barat, 2000)

Table 8.1: Main difference of HPTLC and TLC - Particle and pore size of sorbents

	HPTLC	TLC
Layer of Sorbent	100µm	250µm
Efficiency	High due to smaller particle size generated	Less
Separations	3 - 5 cm	10 - 15 cm
Analysis Time	Shorter migration distance and the analysis time is greatly reduced	Slower
Solid support	Wide choice of stationary phases like silica gel for normal phase and C_8, C_{18} for reversed phase modes	Silica gel, Alumina & Kiesulguhr
Development chamber	New type that require less amount of mobile phase	More amount
Sample spotting	Auto sampler	Manual spotting
Scanning	Use of UV/ Visible/ Fluorescence scanner scans the entire chromatogram qualitatively and quantitatively and the scanner is an advanced type of densitometer	Not possible

(Source: Poole , 1985)

8.1.1 Steps involved in HPTLC

Below given steps are usually performed in case of manually prepared plates

1. Selection of chromatographic layer
2. Sample and standard preparation
3. Layer pre-washing
4. Layer pre-conditioning
5. Application of sample and standard
6. Chromatographic development
7. Detection of spots
8. Scanning
9. Documentation of chromatic plate

8.1.2 Difference between HPTLC and HPLC/TLC

(i) Three times more samples can be applied on a plate

(ii) The separation is affected 4-6 times faster

(iii) Unlike HPLC, the stationary phase in HPTLC can be preserved
(iv) Large sample volume can be applied by spray on a technique which eliminates a pre concentration step
(v) A large number of samples are chromatographed side by side
(vi) Any solvent can be used

Comparison between TLC and HPTLC is given in Table 8.2. In HPTLC, pre coated aluminium sheets or glass plates are used, which are readily available commercially, as compared to manually prepared plates.Smaller particle size and high packing density provide better surface homogeneity of HPTLC plates hence large number of spots can be applied on plates and more number of components can be separated. Silca gel is the most commonly used adsorbent for HPTLC plates. However, for specilized applications, HPTLC plates with modifeid layers like RP-18, cyano, diol, amino, chiral etc. are also available. Sample clean-up and extraction processes in HPTLC is minimal that result in considerable saving of time and cost. The following references will be useful for understanding HPTLC in detail: Stahl (1969), Sherma and Zweig (1971), Zlatkis and Kaiser (1977), Kirchner (1978), Bertsch et al., (1980) Touchstone and Dobbins (1983), Poole and Poole (1994) and Handa et al.,(1999).

Table 8.2: Comparison between TLC and HPTLC

Sr.No.	Parameter	TLC	HPTLC
1.	Sample volume	1-5μl	0.1-0.2μl
2.	Spot diameter	3-6mm	1-2.5mm
3.	Diameter of separated spot	6-15mm	2-5mm
4.	Solvent migration	10-15cm	3-6cm
5.	Separation time	30-200min	3-20min
6.	Plate height	20cm	5cm
7.	Effetive theoretical plates	600	5000
8.	Number of spots	10	10-20
9.	Silica gel (Size)	12μm	7μm

8.2 Sample Application

Assuming that the mobile phase and the developmental methods are optimum, the results of TLC depend on the manner in which the sample is applied on to the plate. For spotting the sample on the plate, the sample should be dissolved in a volatile solvent preferably of low polarity. An optimized dosage is achieved by:

(i) using as small a dosage volume as possible or
(ii) using highly concentrated solutions

HPTLC plates require less than 1mm spots for the resolution.This can be accomplished in several ways:

(i) using self loading capillaries

(ii) contact spotting by Camag Nanomet having platinum iridium (Pt-Ir) capillary pipettes in 100-200 μl sizes. Now a days, totally automatic pipets are available

8.2.1 Contact Spotting

For contact spotting, a specially treated flouro polymer film is placed on a perforated plate on a device that can apply either suction or pressure on the film. Sample solutions (50-100 μl) are deposited in depressions formed on a polymer film producing a symmetrical droplet.The solvent is evaporated by gently heated with a hot nitrogen draft on the sample. The residue is transferred to the layer by bringing it in contact with the film and up to 15 samples can be spotted simultaneously using an automated contact spotting device.

8.3 Development of HPTLC Plates

8.3.1 Linear Developement

Normally, TLC is developed in the linear mode, either in ascending, decending or horizontal fashion. With horizontal linear development chambers, development of up to 32 samples on 10 × 10 cm HPTLC plate can take place simultaneously.

8.3.2 Circular Development

Circular development was widely used in early days in all forms of planar chromatography but is replaced by more convenient linear techniques. It is now receiving attention with the development of U-chamber by CAMAG. HPTLC plates can be developed in a circular rather than linear mode by using this chamber.HPTLC plate is placed face down in the holder ring of the body, mobile phase is fed to the centre of the plates through a Pt-Ir capillary and its flow is cotrolled electronically. The layer may be conditioned by gas fed through a circular channel and excited through a bore, the direction of the gas can be reversed. Circular development technique claim many advantages such as increase separating region and thus the separation number, incresed resolution and reproducibility incresed sensitivity and accuracy in quantitation and enhanced speed of analysis.Compounds with lower Rf values are better resolved by circular development than by linear development and is faster than linear HPTLC.

8.3.3 Anticircular Development

Samples are applied on a circle of diammeter slightly less than that of solvent feeding line on the periphery of the circular plate and the solvent is allowed to flow to the centre. The solvent moves inward and converges at the centre of circle. Anticircular TLC has very fast speed and form more compact bands for substances in the high Rf region.

8.3.4 Continuous Development

Continuous developement is useful for separation of strongly retained solutes without changing the solvent or removing the plate from the chamber. One of the greatest advantages of the descending mode of development is its easy adaptability to continuous development. The top end of the plates is extended out of the chamber so that solvent evaporates and its flow is continuous. To facilitate easy evaporation of the solvent, hot air or heat is applied on the edge. Weak solvents are used to increase selectivity and development distances are kept short so that time does not become excessive.

8.3.5 Two Dimensional HPTLC

In a complex mixture of compounds, it is possible that certain components separate well on a particular system and others on another. Two dimensional chromatography may be used to advantage in such cases. Samples spotted in a corner of a plate and developed with two mobile phases of different types at right angles with drying between runs permits separation of complex mixtures. The technique is very versatile and can be modified in a number of ways using different stationary phases and separation techniques.

HPTLC is a completely visual technique and for better results, separated components are sometimes subjected to post chromatographic derivatisation which carried out in a automated dip chambers and plates are uniformly exposed to the derivatising agent leading to accurate and reproducible results. A wide variety of chemical reagents are shown in Table 8.3 for different classes of pesticides.

Table 8.3: Spraying reagents for pesticides

Sr. No.	Organochlorines	Organophosphoros	Fungicides	Pyrethroids	Carbamate
1.	Diphenyl amine	p- dimethylbenzaldehyde	Sodium chlorate	5%ninhydrinin 1-butanol	Dansyl chloride
2.	N,N-dimethyl benzidine	Palladium chloride	Silver nitrate	Ferric nitrate in nitric acid	Phenyl hydrazine hydro chloride
3.	Silver nitrate-Ammonium hydroxide	Silver nitrate	Fluorescamine	5% silver nitrate in nitric acid	p-nitroaniline
4.	N,N-tetramethyl benzidine	Iodine		2% triphenyl-tetrazolium chloride in 1N sodium hydroxide-Methanol	Sodium nitrite
5.	Silver nitrate	Rhodamine B			Toluidine
6.	3-methyl benzidine	p- dimethylamino benzaldehyde-AcOH			Sulfanilic acid
7.	3,5-3',5'-tetramethyl benzidine				Potassium hydroxide

8.4 Automated Multiple Development (AMD) of Thin Layer Chromatograms

Although the multiple developing techniques are not new, it has been used most effectively with HPTLC plates. In this system, the chromatogram is developed repeatedly in the same direction. Each partial run goes over a longer solvent migration distance than the one before (Handa et al., 1999).In this technique, programmer is used to control both the distance and the number of developments. Multiple developments are accomplished with one or more developing solvents. After each development the plate is removed from the chamber, dried, and then developed again. Every subsequent development sweeps the trailing edge of the spot closer to the leading edge. This effectively forces each spot into narrow bands, thus increasing the efficiency of the system. The selection of different mobile phases, coupled with improved efficiency gives the chemist a very powerful tool for separating complex mixtures. The elution strength of the mobile phase is reduced after each successive run (e.g. starting with methanol and progressively increasing the proportion of chloroform and then benzene).With conventional plates, multiple developments can be very time consuming. However, because of the shorter development distances needed with the high performance plates, this is no longer a problem. The combination of focusing effect and gradient elution results in extremely narrow bands. Their typical peak width is about 1 mm. In the AMD system, the Rf is independent of matrix and one can use time outside normal working hours, since the system is fully automated.

8.5 Quantification

Pesticides separated by HPTLC are quantified by *in situ* measurement of absorbed visible, UV or emitted fluorescence light. Sample and standard are chromatographed on same plate and after development, chromatogram is scanned. Scanners employ a photomultiplier detector to measure either reflected or transmitted light, with either single beam, double beam, or single beam dual wavelength optical system. Camag TLC scanner III can be used to scan the chromatogram in reflectance or in transmittance mode by absorbance or by fluorescent mode in which scanning speed is selectable up to 100 mm/s and spectra is recorded. About 36 tracks with up to 100 peak windows can be evaluated. With good quality HPTLC plates and regularly shaped and well resolved zones, excellent quantitative results are obtained. The ability to spot unknown samples and standards on the same plate and to subject them to the same chromatographic conditions is an inherent advantage of quantitative HPTLC compared to other chromatographic methods. Systematic errors are minimized and accuracy and precision values compare favourably to other methods of analysis. Concentration of analyte in the sample is calculated by considering the sample initially taken and dilution factors.

8.6 Application of HPTLC Technique in the Analysis of Pesticides

Jork and Roth (1977) separated 14 triazine herbicides by HPTLC and sensitivity of the method was 3-5 ng. Sackett (1984) employed HPTLC to assay gibberellins in fermentation broth up to 200 ppm. Hausk and Amadori (1980) studied the separation and quantification of ioxynil and MCPA on HPTLC plate with and without pre adsorbent zone and on high performance cellulose plate. Hausk and Amadori (1980) separated five BHC isomers on silica gel 60 HPTLC plates using toluidine as the chromogenic reagent. Mouratidis and Their (1995) found HPTLC is an effective method for confirming gas-chromatographic results for insecticide and fungicide residues in sample solutions obtained by clean-up of fruit and vegetable extracts according to the Deutsche Forschungsgemeinschaft multiresidue method. The routine limit of determination was 0.01-0.05 mg/kg in most cases. Fan et al., (2007) used high-performance thin layer chromatographic method for analysis of the residues of tricyclazole, thiram, and folpet in tomatoes by extracting with acetone-dichloromethane 1:1 (v/v) by mechanical vibration. The plates were developed with hexane-acetone 6+4 (v/v) in an unsaturated glass twin-trough chamber and developed HPTLC plates densitometrically showing detection limits of tricyclazole ($R_F = 0.26$), thiram ($R_F = 0.65$), and folpet ($R_F = 0.77$) were 1.2×10^{-8} , 3.0×10^{-8} , 4.0×10^{-8} g, respectively. Recoveries of the pesticides from tomatoes with this analytical method were 67.66–98.02%, and RSD were 0.13–22.06%. For the determination of chlorpyriphos, acephate, dichlorovos, carbofuran and imidachloprid residues in brinjal fruits, Iqbal et al., (2007) used a HPTLC method. Brinjal samples were extracted by ethyl acetate and analyzed by HPTLC with enzyme inhibition horse blood serum method (acetylcholine esterase enzyme) which was very sensitive for the detection of insecticide residues. The extract was spotted on silica gel plate, which was developed in mobile phase (ethyl acetate) and spot visibility was determined after spraying with acetylcholine esterase enzyme and tris-buffer solution. Rf value and average spot diameter was measured and imidachloprid, carbofuran, acephate, dichlorovos and chlorpyriphos residue concentration in ppm.

8.6.1 Applications of HPTLC

HPTLC can be used for analysis in:

(i) Food chemistry e.g. carbohydrates, aflatoxins and residue analysis
(ii) Environmental chemistry e.g. pollutants and pesticide residues
(iii) Clinical chemistry e.g. steroids, therapeutic drug monitoring
(iv) Pharmaceutical industry e.g. stability tests and pharmacokinetics
(v) Forensic laboratories e.g. drugs, screening, poisons and pollutants
(vi) Efficient and cheap technique for agrochemicals

References

Barat, S.,2000. HPTLC- An important analytical tool for pesticides residue analysis. In: *Pesticide residue analysis* (eds. Yadav, P.R., Kathpal, T.S., and Rohilla, H.R.), Department of Entomology, CCSHAU, Hisar:103-107.

Bertsch, W., Hara, S., Kaiser, R.E., and Zlatkis, A., 1980. *Instrumental HPTLC*, Huethig, Heidelberg,Germany.

Fan, W., Yue, Y ., Tang, F., Cao , H., 2007. Use of HPTLC for simultaneous determination of three fungicides in tomatoes. *JPC - Journal of Planar Chromatography - Modern TLC*, 20(6): 419-421.

Hausk, H.E., and Amadori, E., 1980. Pesticide analytical methodology (eds. Harvey J. and Zweig , G.): 159 (ACS series no.136).

Iqbal, M. F., Maqbool, U., Asi, M. R., and Aslam, S., 2007. Determination of pesticide residues in brinjal fruit at supervised trial. *J. Anim. Pl. Sci.* 17(1-2):21-23.

Jork, H., and Roth, B.,1977. Vergleichende chromatographische untersuchungen bei s-triazinen. *J. Chromatographia.* 144:39-56.

Kirchner, J.G., 1978. *Thin-Layer Chromatography*, 2nd edn., Wiley, New York, USA.

Mouratidis, S., and Their, H.P., 1995. Solid phase extraction for the confirmation of results in polar pesticides residue analysis by HPTLC. *Zeitschrift für Lebensmitteluntersuchung und -Forschung A.* 201 (4):327-330.

Poole, C.F., 1985. *Anal.Chem.* 4:209.

Poole, C.F., Poole, S.K., 1994. Instrumental thin-layer chromatography. *Anal. Chem.* 66 (1): 27A–37A.

Sackett, P.H., 1984. High-performance thin-layer chromatography of gibberellins in fermentation broths. *Anal.Chem.*56:1600-1603.

Sherma, J., and Zweig, G., 1971. *Paper Chromatography and Electrophoresis:* Vol. II. Paper Chromatography, Academic Press, New York, USA.

Stahl, H., 1969. *Thin–Layer Chromatography: A Laboratory Handbook.* Springer-Verlag, Berlin, Germany.

Touchstone, J.C., and Dobbins, M.F., 1983. *Practice of Thin-Layer Chromatography.* 2nd edn., Wiley, New York, USA.

Zlatkis, A., and Kaiser, R.E., 1977. In: HPTLC: High Performance Thin Layer Chromatography. (eds. Zlatkis, A. Kaiser, R.E.) Elsevier, Amsterdam, Netherlands:12.

❑❑❑

Chapter 9
Maximum Residue Limits (MRL)

Crops treated with pesticides invariably contain small amount of these chemicals and the hazard depends on the amount of pesticide residues that remain on the crop and their toxicity. The amount of residue that may remain on crop or food commodity and environmental commodities depend on the nature of pesticides, good agricultural practice, environmental conditions and adoption of safe waiting periods and other perfect methods of storage. As residues are inevitable by product of pesticide use, hence pesticide residue in food commodities and their entry in to food chain has become major cause of concern all over the world. The plant science industry has a responsibility to assist farmers to produce crops that are acceptable to the food chain and consumers alike.

It is clear that pesticide residues in farm produce should be kept as low as practicable, and eliminated whenever possible through Good Agricultural Practices (GAP). However, the use of pesticides can be regulated to ensure minimum residues on food which can be considered safe to human consumption and environment as a whole.

To regulate pesticide residues to safe levels, a concept was introduced by Joint FAO/WHO Expert Committee on Food Additives (1955) and *Codex Alimentarius Commission* was established in 1964 to implement the Joint FAO/WHO Food Standard Programmes. The Commission has laid down principles for aiming at the *maximum residue limits* (Tolerance limits) of pesticide on food commodities. The recommendations are based on:

(i) reports of the supervised trials designed to determine the maximum residues likely to occur on food through the application of minimum effective dosage of pesticide to control pest

(ii) detailed toxicological studies with parent compound and its major metabolites to determine possible adverse effect on animals

9.1 Maximum Residue Limit (MRL)

Maximum residue limit is the maximum concentration for a pesticide residue (on crop or food commodity resulting from the use of pesticides according to good agricultural practice. The concentration is expressed in milligrams of pesticide residue per kilogram of the commodity. Maximum residue levels are set on the basis of supervised trials in which good agricultural practice (GAP) is observed and must not pose an unacceptable risk to human health.

or

Tolerance

A pesticide tolerance is the maximum amount of a pesticide residue that can be present on food. The tolerance is expressed in ppm, or parts of the pesticide per million parts of food by weight. A chemical generally regarded as safe need not be given a tolerance, or the chemical may be exempted from the requirement for a tolerance. This might occur when the chemical is used on non bearing fruit trees, for example. For toxic chemicals, unified standards need to be set for raw and processed foods.

9.1.1 The Basis for Tolerance

The process of setting a tolerance level on food, first requires the determination of a no observable-effect level, or NOEL. A NOEL is the daily dosage of a pesticide that, when administered over a long period, produces no observable symptoms or pathological effects in experimental animals. A fraction of that dose, usually one-hundredth, is then set as an acceptable daily intake (ADI) for humans.A major problem in establishing a tolerance is how to extrapolate from toxicity data for experimental animals to the probable safety level of the pesticide in humans. This is handled by assuming that humans are 10 times as sensitive to the pesticide as the rat and that their range to such substances could vary by a factor of 10. Thus, a 100-fold margin of safety is introduced ($10 \times 10 = 100$). This means that the NOEL is divided by 100 and the resulting figure is used as the upper limit for the amount allowed in human food. In the case of infants and children, a 1000-fold margin of safety is used. Therefore, the tolerance on each food is set sufficiently low so that the daily consumption of the particular food will not exceed the ADI for the pesticide.

9.2 Acceptable Daily Intake (ADI)

The acceptable daily intake is defined as the amount of a food additive that can be ingested daily in the diet without appreciable risk on the basis of all facts known at the time. "Without appreciable risk" refers to the practical certainty that injury will not result, even after a lifetime of experience. The ADI is usually given in mg per kg body weight per day, (Lu and Kacew 2002, Mackey and Kotsonis 2002). This concept was first introduced in 1961 by the Council of

Europe and later the Joint FAO/WHO Expert Committee on Food Additives (JECFA), a committee maintained by two United Nations bodies: the Food and Agriculture Organization (FAO) and the World Health Organization (WHO) (Lu and Kacew (2002).

9.2.1 Purpose of an ADI

The ADI is a practical approach to determine the safety of food additives and is a means of achieving some uniformity of approach in regulatory control. It serves to ensure that the actual human intake of a substance is well below toxic levels.

9.2.2 Determination of ADI

It is based on a scientific review of all available toxicological data on a specific additive D both observations in humans and tests in animals. Laboratory tests in animals determine the maximum dietary level of the additive that is without demonstrable toxic effects, i.e., the "No Observable Effect Level" (NOEL) Fennema, (1996). This level is then extrapolated to man by dividing the no-effect level by a large factor, often 100. This results in a substantially lower level for man, and thus a large margin of safety.

$$\text{ADI} = \frac{\text{NOAEL}}{10}$$

9.3 No Observable Adverse Effect Level (NOAEL)

The NOAEL is the highest dose of a substance in experimental animal studies that does not cause any detectable toxic effects. The NOAEL is expressed in milligrams of the substance per kilogram of body weight per day.NOAEL (No observable adverse effect level) denotes the level of exposure of an organism, found by experiment or observation, at which there is no biologically or statistically significant (e.g. alteration of morphology, functional capacity, growth, development or life span) increase in the frequency or severity of any adverse effects in the exposed population when compared to its appropriate control.

9.4 Maximum Permissible Intake (MPI)

MPI = Acceptable daily intake × Average body weight

9.4.1 Relationship between MRL and ADI

MRL are proposed only for pesticides for which ADI has been established. Apart from this MRL and ADI have no direct relationship. They are based on separate evaluations of different kinds of data and describe two totally different figures, namely,

(i) the maximum concentration of pesticide residue that might be expected to result from good agricultural practice

(ii) the level of residue that if ingested daily would be without risk to the consumer

9.5 Fixation of Maximum Residue Limits at the International Level

The MRL is set by national governments on a conservative principle to ensure the farmer uses as little as possible of the chemical to achieve the desired result on the pest, a type of precautionary approach to minimise any potentially harmful effects. The MRL is not determined directly from the ADI. MRLs are set for trading purposes, they are not safety levels. Different countries have different trading levels, but generally do not have different safety standards.

At the international level, the maximum residue limits of pesticides are fixed by FAO/WHO committee on Pesticide Residues (Handa et al., 1999). The Codex Committee on Pesticide Residues (CCPR) is an inter-governmental body which advises the Codex Alimentarius Commission. The FAO/WHO Alimentarius Commission was established to implement the Joint FAO/WHO Food Standards Programme and have 183 (182 member countries and one member organisation) members of the Commission. The Codex Committee on Food Standards has the responsibility to:

(i) establish maximum limits for pesticide residues in food items or in groups in food.

(ii) establish maximum limits for pesticide residues in certain animal feeding stuffs moving in international trade where this is justified for reasons of protection of human health.

(iii) consider methods of sampling and analysis for the determination of pesticide residues in food and feed.

(iv) consider other matters in relation to the safety of food and feed containing pesticide residues.

(v) establish maximum limits for environmental and industrial contaminants showing chemical or other similarity to pesticides in specific food items or groups of food.

9.6 Fixation of Maximum Residues Limits in India

In India the maximum residues limits (MRL) are fixed by Central Codex Committee of Food Standards, a Statutory Committee under Prevention and Adulteration Act (PFA), with the Plant Protection Adviser to the Government of India as its Chairman. Codex India the National Codex Contact Point (NCCP) for India, is located at the Directorate General of Health Services, Ministry of Health and Family Welfare (MOH&FW), Government of India. It coordinates and promotes Codex activities in India in association with the National Codex Committee and facilitates India's input to the work of Codex through an established consultation process. The responsibility of enforcement of maximum residue limit lies with State governments. MRL values fixed for different pesticides in various food commodities is given in Table 9.1.

Table 9.1: Maximum Residue Limits as per Prevention of Food Adulteration Act, 1954 (up dated 2009)

Sr. No.	Name of insecticide	Food	Tolerance limit mg/kg (ppm)
1.	Acephate	Safflower seed	2.00
		Cottonseed	2.00
2.	Acetamiprid	Cotton seed oil	0.10
3.	Alachlor	Cotton seed	0.05
		Groundnut	0.05
		Maize	0.10
		Soybeans	0.50
4.	Aldicarb (sum of aldicarb, its sulphoxide and sulphone, expressed as aldicarb)	Potato	0.50
		Chewing Tobacco	0.10
5.	Aldrin, dieldrin (The limits apply to aldrin and dieldrin singly or in any combination and are expressed as dieldrin)	Food grains	0.01
		Milled food grains	NIL
		Milk and milk products (fat basis)	0.15
		Fruits and Vegetables	0.10
		Meat	0.20
		Eggs (shell-free basis)	0.10
6.	Alfa Naphthyl Acetic Acid (A.N.A.)	Pineapple	0.50
7.	Anilophos	Rice	0.10
8.	Atrazine	Maize	NIL
		Sugarcane	0.25
9.	Azoxystrobin	Tomato	1.0
		Mango	0.01
		Chilli	1.0
10.	Benomyl	Food grains	0.50
		Milled food grains	0.12
		Vegetables	0.50
		Mango	2.00
		Banana (whole)	1.00
		Other fruits	5.00
		Cotton seed	0.10
		Groundnut	0.10
		Sugar beet	0.10
		Dry fruits	0.10
		Eggs (shell free basis)	0.10
		Meat and Poultry(carcass fat basis)	0.10
		Milk and Milk products(fat basis)	0.10
11.	Bitertanol	Wheat	0.05
		Groundnut	0.10

12.	Butachlor	Rice	0.05
13.	Bifenthrin	Cotton Seed	0.05
14.	Benfuracarb	Red Gram	0.05
		Rice	0.05
15.	Buprofezin	Rice	0.05
16.	Bensulfuron methyl	Rice	0.01
17.	Carbosulfan	Rice	0.20
18.	Carbendazim	Food grains	0.50
		Milled food grains	0.12
		Vegetables	0.50
		Mango	2.00
		Banana (whole)	1.00
		Other fruits	5.00
		Cotton seed	0.10
		Groundnut	0.10
		Sugar beet	0.10
		Dry fruits	0.10
		Eggs (shell free basis)	0.10
		Meat and Poultry(carcass fat basis)	0.10
		Milk and Milk products(fat basis)	0.10
19.	Carbaryl	Fish	0.20
		Food grains	1.50
		Milled food grains	NIL
		Okra and leafy vegetables	10.00
		Potatoes	0.20
		Other vegetables	5.00
		Cottonseed (whole)	1.00
		Maize cob (kernels)	1.00
		Maize	0.50
		Rice	2.50
		Chillies	5.00
20.	Chlorimuron-ethyl	Wheat	0.05
21.	Chlordane (residue to be measured as cis plus trans chlordane)	Food grains	0.02
		Milled food grains	NIL
		Milk and milk products(fat basis)	0.05
		Vegetables	0.20
		Fruits	0.10
		Sugar (beet)	0.30
22.	Chlorfenvinphos (Residues to be measured as α and β isomers of chlorfenvinphos)	Food grains	0.025
		Milled Food grains	0.006
		Milk and Milk Products(fat basis)	0.20
		Meat and Poultry(carcass fat)	0.20
		Vegetables	0.05

		Groundnuts(shell free basis)	0.05
		Cotton seed	0.05
23.	Chlorobenzilate	Fruit	1.00
		Dry Fruits, Almond and Walnuts (shell free basis)	0.20
24.	Chlorpyrifos	Food grains	0.05
		Milled food grains	0.01
		Fruits	0.50
		Potatoes and onions	0.01
		Cauliflower and cabbage	0.01
		Other vegetables	0.20
		Meat & Poultry(carcass fat basis)	0.10
		Milk & Milk Products(fat basis)	0.01
		Cotton seed	0.05
		Cotton seed oil (crude)	0.025
25.	Captan	Fruit and Vegetable	15.00
26.	Carbofuran (sum of carbofuran and 3-hydroxy carbofuran expressed as carbofuran)	Food grains	0.10
		Milled food grains	0.03
		Fruit and Vegetable	0.10
		Oil seeds	0.10
		Sugarcane	0.10
		Meat and Poultry(carcass fat basis)	0.10
		Milk and Milk products(fat basis)	0.05
27.	Copper oxychloride (determined as copper)	Fruit	20.00
		Potato	1.00
		Other vegetables	20.00
28.	Cypermethrin (sum of isomers) (fat soluble residue)	Wheat grains	0.05
		Milled wheat grains	0.01
		Brinjal	0.20
		Cabbage	2.00
		Bhindi	0.20
		Oil seeds except groundnut	0.20
		Meat and Poultry(carcass fat basis)	0.20
		Milk and Milk products(fat basis)	0.01
29.	Captafol	Tomato	5.00
30.	Cartaphydrochloride	Rice	0.50
31.	Chlormequatchloride	Grape	1.00
		Cotton seed	1.00
32.	Chlorothalonil	Groundnut	0.10
		Potato	0.10
33.	Cyhalofop-butyl	Rice	0.50
34.	Clodinafop-propargyl	Wheat	0.10
35.	Cymoxanil	Grapes	0.10

36.	Clomazone	Rice	0.10
		Soybean Seed	0.10
		Soybean Seed Oil	0.10
37.	β-cyfluthrin	Cotton Seed	0.02
38.	Chlorfenopyr	Cabbage	0.05
		Chilli (Green)	0.05
39.	Chlorantraniliprole	Rice	0.03
		Cabbage	0.03
		Sugarcane	0.03
		Cotton	0.03
40.	Carpropamid	Rice	1.0
41.	D.D.T. (The limits apply to D.D.T.,D.D.D. and D.D.E. singly or in any combination)	Milk and milk products(fat basis)	1.25
		Fruits and Vegetables including potatoes	3.50
		Meat, poultry and fish(whole basis)	7.00
		Eggs (shell free basis)	0.50
42.	Diazinon	Food grains	0.05
		Milled food grains	NIL
		Vegetables	0.50
43.	Dichlorvos [content of dichloracetaldehyde(D.D.A) be reported where possible]	Food grains	1.00
		Milled food grains	0.25
		Vegetables	0.15
		Fruits	0.10
44.	Dicofol	Fruits and Vegetables	5.00
		Tea (dry manufactured)	5.00
		Chillies	1.00
45.	Dimethoate (residue to be determined as dimethoate and expressed as dimethoate	Fruits and Vegetables	2.00
		Chillies	0.50
46.	Decamethrin/Deltamethrin	Cotton seed	0.10
		Food grains	0.50
		Milled food grains	0.20
47.	Dithiocarbamates (the residue tolerance limit are determined and expressed at mg/CS_2/kg and referred separately to the residues arising from any or each groups of dithiocarbamates (a) Dimethyl dithiocar- bamates residue resulting from the use of ferbam or Ziram,	Food grains	0.20
		Milled food grains	0.05
		Potatoes	0.10
		Tomatoes	3.00
		Cherries	1.00
		Other fruits	3.00

	(b) Ethylene bis - dithiocarbamates resulting from the use of Mancozeb maneb or Zineb (including zineb derived from nabam plus zinc sulphate), and (c) Mancozeb	Chillies	1.00
48.	Diflubenzuron	Cotton seed	0.20
49.	Dodine	Apple	5.00
50.	Diuron	Cotton seed	1.00
		Banana	0.10
		Maize	0.50
		Ciytud (Sweet Orange)	1.00
		Grapes	1.00
51.	Diclofop-methyl	Wheat	0.10
52.	Dithianon	Apple	0.10
53.	2,4-D	Food grains	0.01
		Milled food grains	0.003
		Potatoes	0.20
		Milk and Milk products	0.05
		Meat and Poultry	0.05
		Eggs (shell free basis)	0.05
		Fruits	2.00
54.	Difenoconazole	Apple	0.01
55.	Dimethomorph	Grapes	0.05
		Potatoes	0.05
56.	Diafenthiuron	Cotton Seed Oil	1.0
		Cabbage	1.0
57.	Endosulfan (residues are measured and reported as total of Endosulfan A and B and Endosulfan-sulphate)	Fruits and Vegetables	2.00
		Cottonseed	0.50
		Cottonseed oil(crude)	0.20
		Bengal gram	0.20
		Pigeon pea	0.10
		Fish	0.20
		Chillies	1.00
		Cardamom	1.00
58.	Ethion (Residues to be determined as ethion and its oxygen analogueand expressed as ethion	Tea (dry manufactured)	5.00
		Cucumber and Squash	0.50
		Other vegetables	1.00
		Cotton seed	0.50
		Milk and Milk Products(fat basis)	0.50
		Meat and Poultry(carcass fat basis)	0.20
		Eggs (shell free basis)	0.20
		Food grains	0.025
		Milled food grains	0.006

		Peaches	1.00
		Other fruits	2.00
		Dry fruits(shell free basis)	0.10
59.	Edifenphos	Rice	0.02
		Rice bran	1.00
		Eggs (shell free basis)	0.01
		Meat and Poultry(carcass fat basis)	0.02
		Milk and Milk products(fat basis)	0.01
60.	Ethephon	Pineapple	2.00
		Coffee	0.10
		Tomato	2.00
		Mango	2.00
61.	Ethoxysulfuron	Rice	0.01
62.	Ethofenprox	Rice	0.01
63.	Fenitrothion	Food grains	0.02
		Milled food grains	0.005
		Milk and Milk Products(fat basis)	0.05
		Fruits	0.50
		Vegetables	0.30
		Meat	0.03
64.	Formothion (determined as and its oxygen analogue and expressed as Dimethoate except in citrus fruits where it is to be determined as formothion)	Citrus fruits	0.20
		Other fruits	1.00
		Vegetables	2.00
		Peppers and Tomatoes	1.00
65.	Fenthion (sum of fenthion, its oxygen analogue and their sulphoxides and sulphones, expressed as fenthion)	Food grains	0.10
		Milled food grains	0.03
		Onion	0.10
		Potatoes	0.05
		Beans	0.10
		Peas	0.50
		Tomatoes	0.50
		Other vegetables	1.00
		Musk melon	2.00
		Meat & Poultry(carcass fat basis)	2.00
		Milk & Milk products(fat basis)	0.05
66.	Fenvalerate (fat soluble residue)	Cauliflower	2.00
		Brinjal	2.00
		Okra	2.00

		Cotton seed	0.20
		Cottonseed oil	0.10
		Meat & Poultry(carcass fat basis)	1.00
		Milk & Milk products(fat basis)	0.01
67.	Fluchloralin	Cotton seed	0.05
		Soybeans	0.05
68.	Fenpropathrin	Cotton seed oil	0.05
69.	Fenarimol	Apple	5.00
70.	Fenazaquin	Tea	3.00
		Apple	0.2
		Chilli (Green)	0.5
71.	Fluvalinate	Cotton seed oil	0.05
72.	Fenobucarb	Rice	0.01
73.	Fenoxy-prop-p-ethyl	Wheat	0.02
		Soybean seed	0.02
74.	Fosetyl-AL	Grapes	10.00
		Cadamom	0.2
75.	Famoxadone	Grapes	0.05
76.	Fenpyroximate	Coconut Water	0.02
		Tea (Black)	0.20
77.	Forchlorfenuron	Grapes	0.01
78.	Fenamidone	Potato	0.01
		Grapes	0.05
79.	Flufenacet	Rice	0.05
80.	Glyphosate	Tea	1.00
81.	Glufosinate-ammonium	Tea	0.01
82.	Heptachlor (combined residues	Food grains	0.01
	of Heptachlor and epoxide to be	Milled food grains	0.002
	determined and expressed as	Milk & Milk products(fat basis)	0.15
	heptachlor)	Vegetables	0.05
83.	Hydrogen cyanide	Food grains	37.5
		Milled food grains	3.00
84.	Hydrogen phosphide	Food grains	NIL
		Milled food grains	NIL
85.	Hexachlorocyclohexane and		
	its isomers		
	(a) Alfa (α) Isomer	Rice grain unpolished	0.10
		Rice grain polished	0.05
		Milk (whole)	0.02
		Fruits and vegetables	1.00
		Fish	0.25
	(b) Beta (β) Isomer	Rice grain unpolished	0.10
		Rice grain polished	0.05
		Milk (whole)	0.02
		Fruits and vegetables	1.00
		Fish	0.25

	(c) Gamma (γ) Isomer: known as Lindane	Food grains except rice	0.10
		Milled food grains	NIL
		Rice grain unpolished	0.10
		Rice grain polished	0.05
		Milk (whole basis)	0.01
		Milk products (having less than 2 per cent fat)(whole basis)	0.20
		Fruits and vegetables	1.00
		Fish	0.25
		Eggs (shell free basis)	0.10
		Meat and poultry(whole basis)	2.00
	(d) Delta (δ) Isomer	Rice grain unpolished	0.10
		Rice grain polished	0.05
		Milk (whole)	0.02
		Fruits and vegetables	1.00
		Fish	0.25
86.	Hexaconazole	Apple	0.10
87.	Hexythiazox	Tea	0.01
		Chilli(Green)	0.01
		Dried Chilli	0.01
88.	Hexazinone	Sugarcane	0.02
89.	Imidacloprid	Cotton seed oil	0.05
		Rice	0.05
90.	Iprodione	Rape seed	0.50
		Mustard	0.50
		Rice	10.0
		Tomato	5.00
		Grapes	10.0
91.	lmazethapyr	Soybean oil	0.10
		Groundnut oil	0.10
92.	Inorganic bromide (determined and expressed as total bromide from all sources)	Food grains	25.00
		Milled food grains	25.00
		Fruits	30.00
		Dried Fruits	30.00
		Spices	400.0
93.	Isoprothiolane	Rice	0.10
94.	Isoproturon	Wheat	0.10
95.	Indoxacarb	Cotton seed	0.1
		Cotton seed oil	0.1
		Cabbage	0.1
96.	Kitazin	Rice	0.20
97.	Kasugamycin	Rice	0.05
		Tomato	0.05
98.	Linuron	Pea	0.05
99.	Lambdacyhalothrin	Cotton seed oil	0.05

100.	Lufenuron	Cabage	0.3
101.	Malathion (Malathion to be determined and expressed as combined residue of malathion and malaoxon	Food grains	4.00
		Milled food grains	1.00
		Fruits	4.00
		Vegetables	3.00
		Dried fruits	8.00
102.	Monocrotophos	Food grains	0.025
		Milled food grains	0.006
		Citrus fruits	0.20
		Other fruits	1.00
		Carrot, Turnip, Potatoes and Sugar beet	0.05
		Onion and peas	0.10
		Other vegetables	0.20
		Cotton seed	0.10
		Cottonseed oil (raw)	0.05
		Chillies	0.20
		Cardamom	0.50
		Meat and Poultry	0.02
		Milk and Milk Products	0.02
		Egg (shell free basis)	0.02
		Coffee (raw beans)	0.10
103.	Malic Hydrazide	Onion	15.00
		Potato	50.00
104.	Metalaxyl	Bajra	0.05
		Maize	0.05
		Sorghum	0.05
105.	Methomyl	Cotton seed	0.10
106.	Methyl Chloro-phenoxy acetic acid (M.C.P.A.)	Rice	0.05
		Wheat	0.05
107.	Myclobutanil	Groundnut seed	0.10
		Grapes	1.00
108.	Metolachlor	Soybean oil	0.05
109.	Metribuzin	Soybean oil	0.10
110.	Metasulfuron methyl	Wheat	0.10
111.	Methbenzthiazuron	Wheat	0.50
112.	Methamidophos(a metabolite of Acephate)	Safflower seed	0.10
		Cotton seed	0.10
113.	Metiram	Tomato	5.0
		Ground nut Seed	0.1
		Ground nut Seed Oil	0.1
114.	Metalaxyl M	Black Pepper	0.5
		Mustard Seed	0.01
115.	Mepiquat Cloride	Cotton Seed	0.5
		Cotton Seed Oil	0.5

116.	Metsulfuron methyl	Rice	0.01
117.	Novaluron	Cotton Seed	0.01
		Cotton Seed Oil	0.01
		Tomato	0.01
		Cabbage	0.01
118.	Oxadiazon	Rice	0.03
119.	Oxydemeton methyl	Food grains	0.02
120.	Oxyflorfen	Rice	0.05
		Groundnut oil	0.05
121.	Oxadiargyl	Rice	0.1
122.	Oxyfllurofen	Teen	0.20
		Potato	0.10
		Onion	0.05
123.	Parathion (combined residues of parathion and paraoxon to be determined and expressed as parathion)	Fruits and vegetables	0.50
124.	Parathion methyl (Combined residues of parathion methyl and its oxygen analogue to be determined and expressed as parathion methyl)	Fruits	0.20
		Vegetables	1.00
125.	Phosphamidon (residues expressed as the sum of phosphamidon and its desethyl derivative)	Food grains	0.05
		Milled food grains	NIL
		Fruits and vegetables	0.20
126.	Pyrethrins (Sum of Pyrethrins I and II and other structurally related insecticidal ingredients of pyrethrum)	Food grains	NIL
		Milled food grains	NIL
		Fruits and vegetables	1.00
127.	Paraquat-dichloride(determined as paraquat cations)	Food grains	0.10
		Milled food grains	0.025
		Potatoes	0.20
		Other vegetables	0.05
		Cotton seed	0.20
		Cottonseed oil (edible refined)	0.05
		Milk (whole)	0.01
		Fruits	0.05
128.	Phosalone	Pears	2.00
		Citrus fruits	1.00
		Other fruits	5.00
		Potatoes	0.10
		Other vegetables	1.00
		Rapeseed/mustard oil (crude)	0.05

129.	Phenthoate	Food grains	0.05
		Milled food grains	0.01
		Oil seeds	0.03
		Edible oils	0.01
		Eggs (shell free basis)	0.05
		Meat and poultry (carcass fat basis)	0.05
		Milk & Milk products(fat basis)	0.01
130.	Phorate (sum of phorate, its oxygen analogue and their sulphoxides and sulphones expressed as phorate)	Food grains	0.05
		Milled food grains	0.01
		Tomatoes	0.10
		Other vegetables	0.05
		Fruits	0.05
		Oil seeds	0.05
		Edible oils	0.03
		Sugarcane	0.05
		Eggs (shell free basis)	0.05
		Milk and milk products(fat basis)	0.05
131.	Pirimiphos-methyl	Rice	0.50
		Food grains except rice	5.00
		Milled food grains except rice	1.00
		Eggs (shell free basis)	0.05
		Meat and Poultry(carcass fat basis)	0.05
		Milk and Milk products(fat basis)	0.05
132.	Permethrin	Cucumber	0.50
		Cotton seed	0.50
		Soybeans	0.50
		Sunflower seed	1.00
133.	Profenophos	Cotton seed oil	0.05
134.	Penconazole	Grapes	0.20
135.	Propiconazole	Wheat	0.05
136.	Pendimethalin	Wheat	0.05
		Rice	0.05
		Soybean oil	0.05
		Cotton seed oil	0.05
137.	Pretilachlor	Rice	0.05
138.	Propineb	Apple	1.0
		Pomegranate	0.5
		Potato	0.5
		Green Chillies	2.0
		Grapes	0.5
139.	Propargite	Tea	10.0
140.	Pyraclostrobin	Tomato	0.01
141.	Pyrazosulfuron ethyl	Rice	0.01
142.	Quinalphos	Rice	0.01
		Pigeon pea	0.01
		Cardamom	0.01
		Tea	0.01

		Fish	0.01
		Chillies	0.20
143.	Quizalofop-ethyl	Soybean Seed	0.05
144.	Quizalofop-p-tefuryl	Soybean Seed	0.02
145.	Simazine	Maize	NIL
		Sugarcane	0.25
146.	Sulfosulfuron	Wheat	0.02
147.	Spinosad	Cotton seed oil	0.02
		Cabbage	0.02
		Cauliflower	0.02
148.	Trichlorfon	Food grains	0.05
		Milled food grains	0.0125
		Sugar beet	0.05
		Fruits and vegetables	0.10
		Oil seeds	0.10
		Edible oil (refined)	0.05
		Meat and Poultry	0.10
		Milk (whole)	0.05
149.	Thiometon (Residues determined is sulfoxide and sulphone expressed as thiometon)	Food grains	0.025
		Milled food grains	0.006
		Fruits	0.50
		Potatoes, Carrots and sugar beets	0.05
		Other vegetables	2.50
150.	Thiophanatemethyl	Apple	5.00
		Papaya	7.00
151.	Triazophos	Chillies	0.20
		Rice	0.05
		Cotton seed oil	0.10
		Soybean oil	0.05
152.	Tridemorph	Wheat	0.10
		Grapes	0.50
		Mango	0.05
153.	Trifluralin	Wheat	0.05
154.	Tricyclazole	Rice	0.02
155.	Triallate	Wheat	0.05
156.	Thiamethoxam	Rice	0.02
157.	Thiodicarb	Cotton	0.02
158.	Triadimefon	Wheat	0.50
		Pea	0.10
		Grapes	2.00
159.	Tebuconazole	Wheat	0.05
160.	Thiocloprid	Cotton Seed	0.05
		Cotton Seed Oil	0.05
		Rice	0.01
161.	Validamycin	Rice	0.01

As per *Ministry of Health and Family Welfare notification G.S.R.* 754 *(E) dt.27-10-2008*

References

Anonymous, 1986. Codex Maximum Limits for Pesticide Residues. *FAO/WHO, Food Standard Programme, CACI Vol. 13. FAO, Rome,Italy.*

Anonymous,1988.Guidelines for Predicting Dietar Intake of Pesticide Residues. *WHO/EHE/FOS/88.2,WHO,Geneva.*

Anonymous, 2009. *Ministry of Health and Family Welfare notification G.S.R. 754 (E)* dt.27-10-2008.

Fennema, O.R., 1996. Food Chemistry. *Marcel Dekker, New York, USA: 828.*

Handa, S.K., Agnihotri, N.P., and Kulshrestha, G., 1999. *Pesticide Residues: Significance, Management and Analysis.* Research Periodicals and Book Publishing House, New Delhi: 13-14.

Lu, F.C., Kacew, S., 2002. Lu's Basic Toxicology: Fundamentals, Target Organs and Risk Assessment. *Taylor & Francis, Washington DC: 364.*

Mackey, M. A., Kotsonis, F.N., 2002. Nutritional Toxicology. Taylor & Francis, Washington DC: 258.

❑❑❑

Chapter 10

Multiresidue Methods for Estimation of Pesticide Residues

In sub-tropical agro-climatic conditions of India, a variety of insect-pest, weeds and diseases hamper the economic growth of field crops. For effective control of these pests, a number of pesticides including insecticides, fungicides, herbicides and nematicides are to be applied. Estimation technique for single pesticide only can not be followed for detection of residues of all categories of pesticides. Hence, this constantly expanding use of pesticides on food crops accentuates the need for rapid, precise and sensitive method for determination of pesticide residues of all the major groups of pesticides. In such situation, multi-residue analytical technique can be efficiently followed for detection and estimation of multiresidues of intra and inter class xenobiotics.

The purpose of multiresidue analyses is to determine the residues of as many pesticides as possible within a short period of time even if the recoveries of some compounds are low. In multiresidue methods the recoveries up to 70% are accepted. The recoveries less than 70% have to be mentioned specifically. Multiresidue analysis is relatively more sensitive and requires extreme vigilance on the part of the analyst during processing of samples. Additionally, instrument, GLC, need to be equipped with capillary columns for high resolution and sophisticated detectors like ECD, NPD and MSD for detecting multi-residues of the pesticides. Multiresidue analytical methods/techniques have been developed for determination of general pesticide residues for some environmental components and different food commodities. Criteria for multi-pesticides screening as per California Department of Food and Agriculture (CDFA) are:

(i) The chemical must be a pesticide.

(ii) The pesticide must be detectable on a representative group of registered food crop.

(iii) Seventy per cent recovery must be obtained unless specifically noted.

(iv) The length of time required for completion of analysis must not be more than 8 hours.

(v) The sensitivity of the analysis must be sufficient to find the chemical at the lowest listed tolerance for fresh food or feed.

(vi) The pesticide analysis must be comparable by at least two analytical techniques (two different columns, two different detectors).

10.1 Multiresidue Methods using GC-Capillary System

For the estimation purpose, multiresidue methods can be classified on the basis of sample (environmental components like soil and water and food commodities like vegetables, milk and milk products, honey, processed food, total diet, etc.) and type of pesticides, organochlorines (OC), synthetic pyrethroids/pyrethroids (SP), organophosphates(OP) and carbamates) in addition to herbicides, fungicides or nematicides. Multiresidue analysis has been divided in to three parts, in Part 1, methods developed by the authors have been explained whereas in Part 2 some latest multiresidues methods developed all over the world for different commodities are explained. Part 3 of this chapter gives berief account of QUECHERS method for multiresidne analysis.

Part A

10.1.1 Estimation of Pesticide Residues in Soil (Kumari et al., 2008)

Method

(i) Take 15g representative air dried and well ground soil in a 100 ml beaker.

(ii) Add two drops (0.25 ml) of ammonia solution (Sp.Gr. 0.91) ad mix properly with glass rod and keep it for half an hour to evaporate off ammonia.

(iii) To the soil add 0.3g activated charcoal, 0.3 g Florisil (pre washed and activated) and 10 g anhydrous sodium sulphate and mix thoroughly.

(iv) Pack the above mixture compactly in a glass column (60 cm × 22 mm, fitted with Teflan stop cock) in between the two layers (2-3 cm) of anhydrous sodium sulphate.

(v) Elute the column with 125 ml acetone: hexane mixture (1:9 v/v) by adjusting the flow rate @ 4 ml/min.

(vi) Extracts can be can be directly used for GLC analysis without further clean-up.

(vii) Concentrate the eluate to near dryness on rotary flash evaporator after adding a drop of mineral oil (liquid paraffin).

(viii) Make the final volume to 2ml in-hexane and inject in GC-ECD system for the estimation of organochlorines and synthetic pyrethroids and GC-NPD capillary system for organophosphates and carbamates.

Flow diagram for extraction and clean-up for multiresidues in soil is given in Figure 10.1

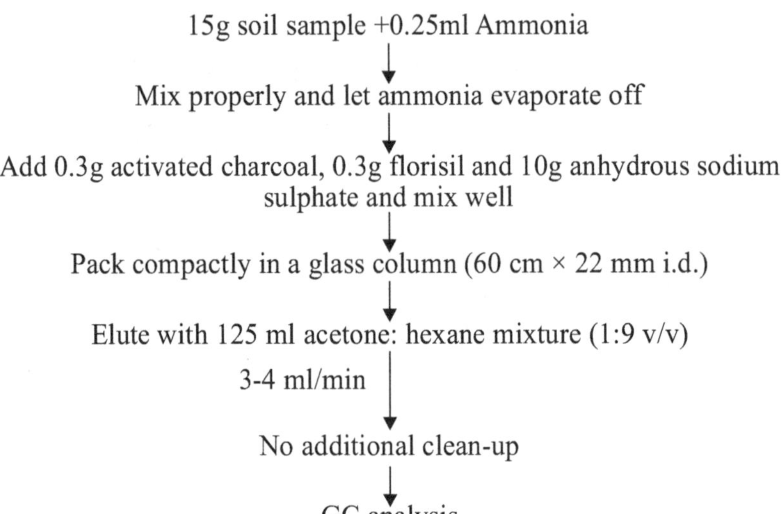

Figure 10.1: Extraction-cum clean-up of OC, OP, SP and carbamates from soil for multiresidues

10.1.2a Estimation of Pesticide Residues in Water (Kumari et al., 2008)

Method

Liquid-liquid extraction (LLE) method used for extraction of pesticide residues from different water samples i.e. drinking water, pond water, river water rain water etc., at ng/L levels (Figure 10.2).

(i) Take 500 ml water in 1L separatory funnel. Add 10-15g NaCl and shake till the salt dissolve completely.

(ii) Shake the funnel gently to dissolve NaCl.

(iii) Add 50 ml 15% dichloromethane (85 ml hexane + 15 ml dichloromethane) in hexane. Shake the funnel vigorously for 1-2 min releasing the pressure intermittently. If emulsions form, first shaking should be slow.

(iv) Keep the funnel undisturbed and allow the layers to separate. Take out the lower aqueous phase in another 1L separatory funnel.

(v) Repeat the process of partitioning two times more using fresh portions of 50 ml of 15% dichloromethane in hexane.

(vi) Combine the upper (dichloromethane/hexane i.e. organic phases).

(vii) Pass the extract/combined organic phase through glass funnel containing 2-3 cm layer of anhydrous sodium sulphate over a small pad of glass wool/cotton at the bottom.

(viii) Concentrate the extract to about 10 ml on rotary flash evaporator at 45-50°C after adding a drop of mineral oil.

(ix) Concentrate the above extract up to near dryness on gas manifold evaporator.

(x) Redissolve the residues in 5 ml hexane.

(xi) Evaporate the contents to near dryness. Repeat the process at least three times more to ensure complete removal of dichloromethane.

(xii) Dissolve the residues in n-hexane for GLC analysis.

(xiii) Recoveries were in the range of 80– 111, 73–95, 83–125 and 82–104 percent for OC, SP, OP insecticides, respectively.

10.1.2b Solid Phase Extraction (SPE)

This method is more sensitive than liquid-liquid extraction. The method is applicable at 1ng/L levels or lowers (Tomkins et al., 1992) (Figure 10.2).

Principle

The water is passed through solid phase extraction cartridge containing octadecyl (C_{18}) groups chemically bonded to silica. Residues retained on extraction medium are eluted with small amount of non polar solvent. Concentration is not required, therefore no evaporative loss.

Method

(i) SPE cartridges are of glass of polypropylene with Teflon frits.

(ii) Pre-wash the cartridge with 10 ml of eluting solvent (acetonitrile/methanol).

(iii) Pass one litre of water through the cartridge at 25-30ml/min.

(iv) Add 10 ml ethyl acetate (eluting solvent) and allow it to pass through cartridge.

(v) Collect the eluate in a 20 ml vial and add one gram anhydrous sodium sulphate to remove excess moisture.

(vi) Extract can be concentrated using manifold evaporator and flowing dry nitrogen.

(vii) Eluate is analysed by GLC using capillary column.

Flow diagram of Liquid-liquid extraction and solid phase extraction are presented in Figure 10.2

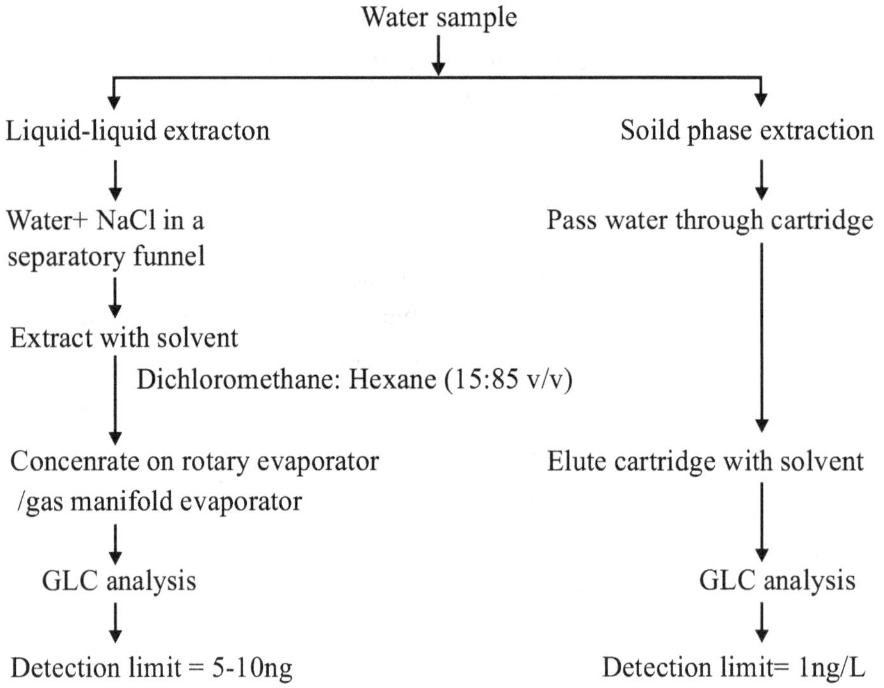

Figure 10.2: Extraction of multiresidues (OC, OP, SP and Carbamates) from water

10.1.3 Estimation of Pesticide Residues in Vegetables and Fruits (Kumari et al., 2001, 2006)

Extraction

(i) Take bulk sample (1-2 kg) of vegetable/fruit.

(ii) Chop it into small pieces and mix properly.

(iii) After quartering take 20g representative sample.

(iv) Macerate it with 4-5g anhydrous sodium sulphate.

(v) Add 100 ml acetone and extract by shaking on mechanical shaker for 1 hour.

(vi) Filter the extract through 2-3 cm layer of anhydrous sodium sulphate.

(vii) Concentrate the extract to 40 ml on rotary flash evaporator after adding a drop of mineral oil.

(viii) Dilute the extract 4-5 times with 10% NaCl aqueous solution.

(ix) Partition it thrice with ethyl acetate (50, 30, 30 ml) in a separatory funnel by shaking vigorously for one minute.

(x) Combine the organic (ethyl acetate) phases and filter through anhydrous sodium sulphate.

(xi) Concentrate the organic phase up to 5 ml on rotary flash evaporator.

(xii) Divide the concentrated extract into two equal parts (one for organochlorines and synthetic pyrethroids and other for organophosphates and carbamates).

Clean-up

For Organochlorines and Synthetic Pyrethroids

(i) Pack the glass column (60 cm × 22 mm i.d) with adsorbent mixture (5g) Florisil: activated charcoal (5:1 w/w) in between two layers of anhydrous sodium sulphate.

(ii) Tap the column gently to ensure uniform and compact packing.

(iii) Prewett the column with 50 ml hexane and transfer the concentrated extract to the column.

(iv) Elute the column with 125 ml solution of ethyl acetate: hexane (3:7 v/v).

(v) Concentrate the eluate to near dryness using rotary flash evaporator followed by gas manifold evaporator after adding one drop of oil.

(vi) Make the final volume to 2 ml in ethyl acetate: n-hexane (3:7 v/v).

For Organophosphates and Carbamates

(i) Pack the glass column (60 x 22 mm i.d.) with adsorbent mixture containing 5g silica gel (60-120 mesh): activated charcoal (5:1 w/w) in between 3-4 cm layers of anhydrous sodium sulphate. Ensure the compact packing of the column by taping gently.

(ii) Prewett the column with 50 ml hexane and load the concentrated extract to the column.

(iii) Elute the column with 125 ml mixture of acetone: hexane (3:7 v/v).

(iv) Concentrate to near dryness using rotary flash evaporator followed by gas manifold evaporator.

(v) Make the final volume to 2 ml in acetone: n-hexane (3:7 v/v).

Flow diagram of extraction of multiresidues from vegetable and fruits is shown in Figure 10.3.

Extraction

Take bulk sample (1-2 kg) of vegetable/fruit

Chop it into small pieces and mix properly

After quartering take 20g representative sample

Macerate it with 4-5g anhydrous sodium sulphate

Add 100 ml acetone and extract by shaking on mechanical shaker for 1 hour

Filter the extract through 2-3 cm layer of anhydrous sodium sulphate

Concentrate the extract to 40 ml on rotary flash evaporator after adding a drop of mineral oil

Dilute the extract 4-5 times with 10% NaCl aqueous solution

Partition it thrice with ethyl acetate (50, 30, 30 ml) in a separatory funnel by shaking vigorously for one minute

Combine the organic (ethyl acetate) phases and filter through anhydrous sodium sulphate

Concentrate the organic phase up to 5 ml on rotary vacuum evaporator

Divide the concentrated extract into two equal parts (one for organochlorines and synthetic pyrethroids and other for organophosphates and carbamates)

Clean-up

For Organochlorines and Synthetic Pyrethroids

Pack the glass column (60 cm × 22 mm i.d) with adsorbent mixture (5g)

Florisil : activated charcoal (5:1 w/w) in between two layers of anhydrous sodium sulphate

↓

Tap the column gently to ensure uniform and compact packing

Prewett the column with 50 ml hexane and transfer the concentrated extract to the column

Elute the column with 125 ml solution of ethyl acetate: hexane (3:7 v/v)

Concentrate the eluate to near dryness using rotary vacuum evaporator followed by gas manifold evaporator after adding one drop of mineral oil

Make the final volume to 2 ml in ethyl acetate: n-hexane (3:7 v/v)

For Organophosphates and Carbamates

Pack the glass column (60 x 22 mm i.d.) with adsorbent mixture containing 5g silica gel (60-120 mesh): activated charcoal (5:1 w/w) in between 3-4 cm layers of anhydrous sodium sulphate

Ensure the compact packing of the column by taping gently

Prewett the column with 50-60 ml hexane and load the concentrated extract to the column

Elute the column with 125 ml mixture of acetone: hexane (3:7 v/v)

Concentrate to near dryness using rotary flash evaporator followed by gas manifold evaporator

Make the final volume to 2 ml in acetone: n-hexane (3:7 v/v)

Figure 10.3: Extraction of multiresidues from vegetables and fruits

10.1.4 Estimation of Pesticide Residues in Jam (Processed food) (Kumari et al., 2008)

(i) Take 5g jam from the homogenized bulk sample.

(ii) Add about 20g anhydrous sodium sulphate and mix thoroughly to make it flowable.

(iii) Add 50 ml acetone and extract it by shaking on mechanical shaker for 1h.
(iv) Filter the extract through 2-3 cm layer of anhydrous sodium sulphate.
(v) Concentrate the extract upto 20 ml on rotary flash evaporator after adding 1 drop of mineral oil.

Clean-up

(i) Add 0.1g activated charcoal to the concentrated extract, shake gently for 30 sec and allow it to settle down.
(ii) Filter the supernatant through Whatman filter paper No.1.
(iii) Repeat the process two times more with 25 ml hexane: acetone (4:1 v/v) mixture.
(iv) Combined the filtrates and add a drop of mineral oil.
(v) Concentrate it to near dryness first on rotary flash evaporator and then on gas manifold evaporator.
(vi) Make the final volume to 2 ml in n-hexane for GC analysis.

10.1.5 Estimation of Pesticide Residues in Milk (Kathpal et al., 2004)

Extraction-cum-Clean-up

(i) Take 10 ml milk from homogenized bulk sample.
(ii) Mix it thoroughly with 10g anhydrous sodium sulphate, 15g silica gel (60-120 mesh) (prewashed first with acetone then with dichloromethane followed by activation at 120^{0}C for 1h) and 5g Florisil (60-100 mesh).
(iii) Pack the mixture in a glass column (60 cm x 22 mm) in between two layers of anhydrous sodium sulphate having a cotton plug at the bottom.
(iv) Tap the column gently to ensure compact and uniform packing.
(v) Elute the column with 150 ml solvent mixture of acetone: dichloromethane (1:1 v/v) at flow rate of 4 ml/min.
(vi) Divide the eluate into two equal parts.
(vii) Add one drop of mineral oil (liquid paraffin) in each part.
(viii) Concentrate one part on rotary flash evaporator followed by manifold evaporator to near dryness.
(ix) Dissolve residues in 5 ml hexane and again evaporate solvent to near dryness on gas manifold evaporator.
(x) Repeat this process 2-3 times to ensure complete removal of dichloromethane traces.
(xi) Make final volume to 2 ml in n-hexane and keep it for analysis of organochlorine and synthetic pyrethroid insecticides.

(xii) Concentrate the second part of eluate to near dryness using rotary flash evaporator and gas manifold evaporator.

(xiii) Make the final volume to 2 ml in ethyl acetate for estimation of organophosphate and carbamate insecticides.

Flow diagram of extraction of multiresidues from milk is shown in Figure10.4.

Extraction-cum-Clean-up

Take 10 ml milk from homogenized bulk sample

↓

Mix it thoroughly with 10g anhydrous sodium sulphate, 15g silica gel (60-120 mesh) (pre washed first with acetone then with dichloromethane followed by activation at 120^0C for 1h) and 5g Florisil (60-100 mesh)

↓

Pack the mixture in a glass column (60 cm x 22 mm) in between two layers of anhydrous sodium sulphate having a cotton plug at the bottom

↓

Tap the column gently to ensure compact and uniform packing.

↓

Elute the column with 150 ml solvent mixture of acetone: dichloromethane (1:1 v/v) at flow rate of 4 ml/min

↓

Divide the eluate into two equal parts

↓

Add one drop of mineral oil in each part

↓

Concentrate one part on rotary vacuum evaporator followed by gas manifold evaporator to near dryness

↓

Dissolve residues in 5 ml hexane and again evaporate solvent to near dryness on gas manifold evaporator

↓

Repeat this process 2-3 times to ensure complete removal of dichloromethane traces

↓

Make final volume to 2 ml in n-hexane and keep it for analysis of organochlorines and synthetic pyrethroids

↓

Concentrate the second part of eluate to near dryness using rotary flash evaporator and gas manifold evaporator

↓

Make the final volume to 2 ml in ethyl acetate for estimation of organophosphates and carbamates

Figure 10.4: Extraction –cum clean-up for multiresidues from milk

10.1.6 Estimation of Pesticide Residues in Feed and Fodder (Kumari et al., 2006)

Extraction

(i) Take 10g representative sample from coarsely powdered bulk sample.

(ii) Add 200 ml of 1% aqueous acetonitrile.

(iii) Extract it for 8 hours on Soxhlet extraction apparatus.

(iv) Transfer the extract to 1L separatory funnel and dilute 4-5 times with 10% NaCl solution.

Liquid-Liquid Partitioning

(i) Partition the extract twice with hexane (2 × 100 ml followed by partitioning twice with dichloromethane (2 × 100 ml) by vigrous shaking for 1 min. each time.

(ii) Combine the both organic phases i.e. of hexane and dichloromethane.

(iii) Add a drop of mineral oil and concentrate to 10 ml on rotary flash evaporator.

Clean up

(i) Pack the glass column (60 cm × 20 mm i.d.) compactly with adsorbent mixture 15g silica gel (60-120 mesh, prewashed and activated at 120°C for 1h), 0.5g activated charcoal and 5g Florisil in between 2-3 cm layers of anhydrous sodium sulphate.

(ii) Prewet the column with 50-60 ml hexane.

(iii) Load the concentrated extract on column and elute with 150 ml mixture of acetone : dichloromethane (1:1 v/v) at a flow rate of 4 ml/min.

(iv) Divide the eluate into two equal portions; one for OC, SP and other for OP and carbamates.

(v) Evaporate first portion to near dryness first on rotary flash evaporator followed by gas manifold evaporator.

(vi) Dissolve the residues in hexane and again concentrate up to dryness.

(vii) Repeat the process three times more to remove traces of dichloromethane.

(viii) Make the final volume to 2 ml in n-hexane for the estimation of organochlorine and synthetic pyrethroid insecticides.

(ix) Evaporate the second portion to near dryness on rotary flash evaporator/gas manifold evaporator.

(x) Make the final volume to 2 ml in ethyl acetate for the estimation of organophosphate and carbamate insecticides

10.1.7 Estimation of Pesticide Residues in Honey (Rathi et al., 1997, Kumari et al., 2003)

Extraction

(i) Take 10g honey from homogenized bulk sample.

(ii) Dissolve it in 100 ml solution of 4% sodium sulphate.

(iii) Transfer it in a 500 ml separatory funnel fitted with teflon stopcock.

(iv) Add 50 ml ethyl acetate and shake the funnel vigrously for 1 min by releasing pressure intermittently.

(v) Collect the upper (ethyl acetate layer) phase.

(vi) Repeat the partitioning two times more using fresh portion of ethylacetate (30 ml) each time.

(vii) Combine all the three portions of organic phases in 200 ml centrifuge tube.

(viii) Centrifuge the extract for 5 minute at 2000 rpm to break the emulsions.

(ix) Take out supernatants and pass through 2-3 cm pad of anhydrous sodium sulphate.

(x) Concentrate it to 5 ml on rotary flash evaporator after adding a drop of mineral oil.

(xi) Concentrated extract will be of light pale colour.

Clean-up

(i) Pack the glass column (60 cm × 22 mm) with mixture of 3g silica gel (60-120 mesh) and 0.5g activated charcoal in between the two layers of anhydrous sodium sulphate.

(ii) Tap the column gently to ensure uniform and compact packing.

(iii) Prewett the column with 50-60 ml hexane.

(iv) Load the concentrated yellow extract to glass column.

(v) Elute it with 100 ml mixture of hexane: acetone (9:1 v/v).

(vi) Concentrate the eluate to near dryness after adding a drop of mineral oil first on rotary flash evaporator followed by gas manifold evaporator.

(vii) Make the final volume to 2 ml in n-hexane.

Flow diagram for extraction of multiresidues in honey is shown below in Figure 10.5.

Extraction

Take 10g honey from homogenized bulk sample

Dissolve it in 100 ml solution of 4% sodium sulphate

Transfer it in a 500 ml separatory funnel fitted with teflon stopcock

Add 50 ml ethyl acetate and shake the funnel vigrously for 1 min by releasing pressure intermittently

Collect the upper (ethyl acetate layer) phase

Repeat the partitioning 2 times more with fresh portion of ethylacetate (30ml) each time

Combine all the three portions of organic phases in 200 ml centrifuge tube

Centrifuge the extract for 5 minute at 2000 rpm to break the emulsions

Take out supernatants and pass through 2-3 cm pad of anhydrous sodium sulphate

Concentrate it to 5 ml on rotary flash evaporator after adding a drop of mineral oil

Concentrated extract will be of light pale colour

Clean-up

Pack the glass column (60 cm × 22 mm) with mixture of 3g silica gel (60-120 mesh) and 0.5g activated charcoal in between the two layers of anhydrous sodium sulphate

Tap the column gently to ensure uniform and compact packing

Prewett the column with 50-60 ml hexane

Load the concentrated extract to glass column

↓

Elute it with 100 ml mixture of hexane: acetone (9:1 v/v)

↓

Concentrate the eluate to near dryness after adding a drop of mineral oil first on rotary vacuum evaporator followed by gas manifold evaporator

Make the final volume to 2 ml in n-hexane.

Figure 10.5: Extraction and clean-up for multiresidues from honey

10.2 GLC Parameters for Estimation of Multiresidues

GLC parameters used for the above developed multiresidue methods are given below:

10.2.1 For Organochlorines (OC) and Synthetic Pyrethroids (SP)

GLC model : Hewlett Packard (HP) 5890A or equivalent

Detector : Electron capture detector (ECD) Ni63

Column : Capillary column, HP-5, 30 m × 0.32 mm i.d. × 0.25 μm film thickness of 5% diphenyl and 95% dimethyl polysilioxane

Temperature Programming

8°/min. 15°min.

Oven 150° → 190° → 280°C

(5 min.) (2 min.) (10 min.)

Injection port temperature : 280°C

Detector temperature : 300°C

Gas flow rate : Total flow 60 ml/min. with split ratio 1:10 (Through column 2ml)

10.2.2 For Organophosphates (OP) and Carbamates

GLC model : HP 5890A or equivalent

Detector : Nitrogen phosphorous detector (NPD)

Column : Capillary column, HP-1, 10 m × 0.53 mm i.d. × 0.65 μm film thickness of 100% diphenyl polysilioxane.

Temperature Programming

10°/Min. 25°min.

Oven 100° → 190° → 240°C

(1 min.) (0 min.) (3 min.)

Injection port temperature :250°C

Detector temperature :275°C

Gas flow rate

H_2 : 1.8 ml/min.

O_2 : 130 ml/min.

N_2 : 18 ml/min

The chromatograms of organochlorines and synthetic pyrethroids are shown in Figure 10.6 whereas of organophosphates and carbamates in Figure 10.7 Retention times of different pesticides observed for above GC conditions are given in Table 10.1

10.2.3 Final Analysis and Calculation

A stock solution of 1-2 ppm analytical grade of standard is injected and retention time is recorded. Then 2-5 μl of cleaned test samples is injected into the GC. The residues are identified by comparing the retention time of sample peaks with that of standard and the amount of residues recorded in the integrator chart. The amount of residues in ppm (μg/g) is calculated as follows:

$$\text{Residues in ppm} = \frac{A_1 \times V_1 \times C}{A_2 \times W}$$

Where,

A_1 = Area of sample in the Chromatogram

A_2 = Area of standard in the Chromatograph

V_1 = Total volume of sample in mL

C = Concentration of analytical standard in ppm (μg/mL)

W = Weight of the sample in g

Rf = Recovery factor

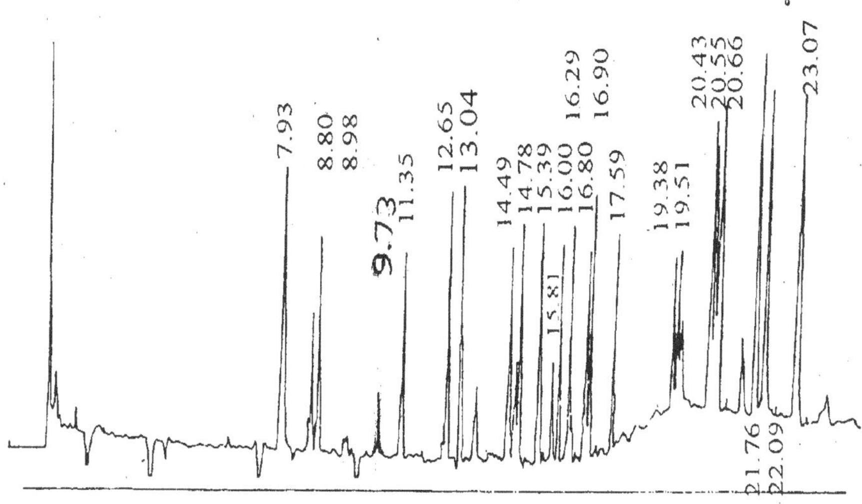

Figure 10.6: GC-ECD Chromatogram of organochlorines and synthetic pyrethroids

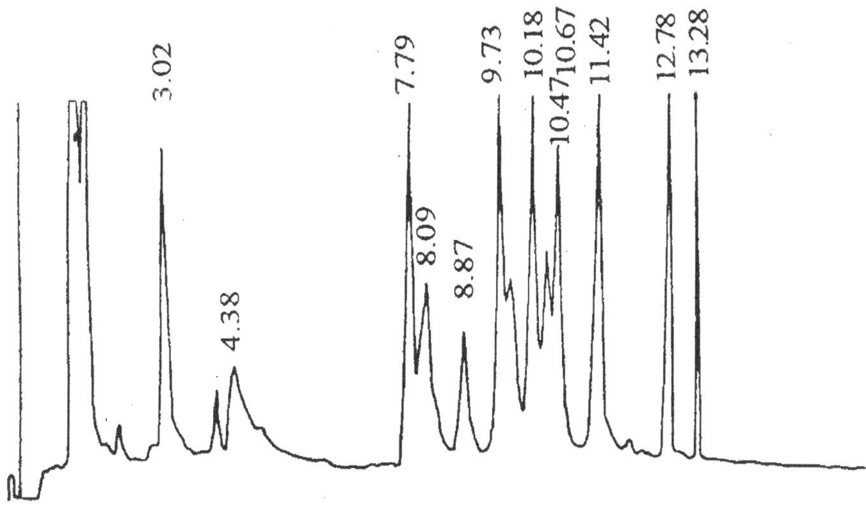

Figure 10.7: GC-NPD Chromatogram of organophosphates and carbamates

Table 10.1: Retention times (Rts) of different pesticides

Sr. No.	Pesticide	Retention time (Rt) (mm)
Organochlorines and Synthetic Pyrethroids		
1.	α-HCH	7.93
2.	β-HCH	8.80
3.	γ-HCH	8.98
4.	δ-HCH	9.73
5.	Heptachlor	11.35
6.	Aldrin	12.65
7.	Chlorpyriphos	13.04
8.	o, p-DDE	14.49
9.	α -Endosulfan	14.78
10.	p, p'-DDE	15.39
11.	p,p'-DDD	15.98
12.	β - Endosulfan	16.00
13.	o, p-DDT	16.29
14.	Endosulfan sulphate	16.80
15.	p, p'-DDT	16.90
17.	Fluvalinate	19.38 19.51
18.	Cypermethrin	20.43 20.55 20.66
19.	Fenvalerate	21.76 22.09
20.	Deltamethrin	23.07
Organophosphates and Carbamates		
21.	DDVP	3.02
22.	Acephate	4.38
23.	Phorate	7.79
24.	Monocrotophos	8.09
25.	Carbofuran	8.87

Contd.

26.	Phosphamidon	9.73
27.	Carbaryl	10.18
28.	Fenitrothion	10.47
29.	Malathion	10.67
30.	Chlorpyriphos	11.42
31.	Quinalphos	12.78
32.	Triazophos	13.28

Part B

10.3 Miscellaneous Multiresidue Methods

10.3.1 Estimation of Multiresidues in Food Grains

Pang et al., (2006) established a method for simultaneous determination of 405 pesticide residues in grain, using accelerated solvent extraction (ASE), solid-phase extraction (SPE), and GC-MS and LC-MS-MS. Samples of grain (10 g) were mixed with Celite 545 (10 g) and the mixture was placed in a 34-mL cell of an accelerated solvent extractor and extracted with acetonitrile in the static state for 3 min with two cycles at 1,500 psig and at 80°C. For the 362 pesticides determined by GC-MS, half of the extracts were cleaned with an Envi-18 cartridge and then further cleaned with Envi-Carb and Sep-Pak NH_2 cartridges in series. The pesticides were eluted with acetonitrile-toluene, 3:1v/v, and the eluates were concentrated and used for analysis after being exchanged with hexane twice. For the 43 pesticides determined by LC-MS-MS the other half of the extracts were cleaned with Sep-Pak Alumina N cartridge and further cleaned with Envi-Carb and Sep-Pak NH_2 cartridges. Pesticides were eluted with acetonitrile-toluene, 3:1. After evaporation to dryness the eluates were diluted with acetonitrile-water, 3:2, and used for analysis. In the linear range of each pesticide the linear correlation coefficient r was equal to or greater than 0.956 and 94% of linear correlation coefficients were greater than 0.990. At low, medium, and high fortification levels, at the limit of detection (LOD), twice the LOD and ten times LOD, respectively, recoveries ranged from 42 to 132%; for 382 pesticides, or 94.32%, recovery was from 60 to 120%. The relative standard deviation (RSD) was always below 38% and was below 30% for 391 pesticides, or 96.54%. The LOD was 0.0005–0.3000 mg kg^{-1}. The proposed method is suitable for determination of 405 pesticide residues in grain such as maize, wheat, oat, rice, and barley, etc.

Flow diagram for extraction of multiresidues in honey is shown below in Figure10.8.

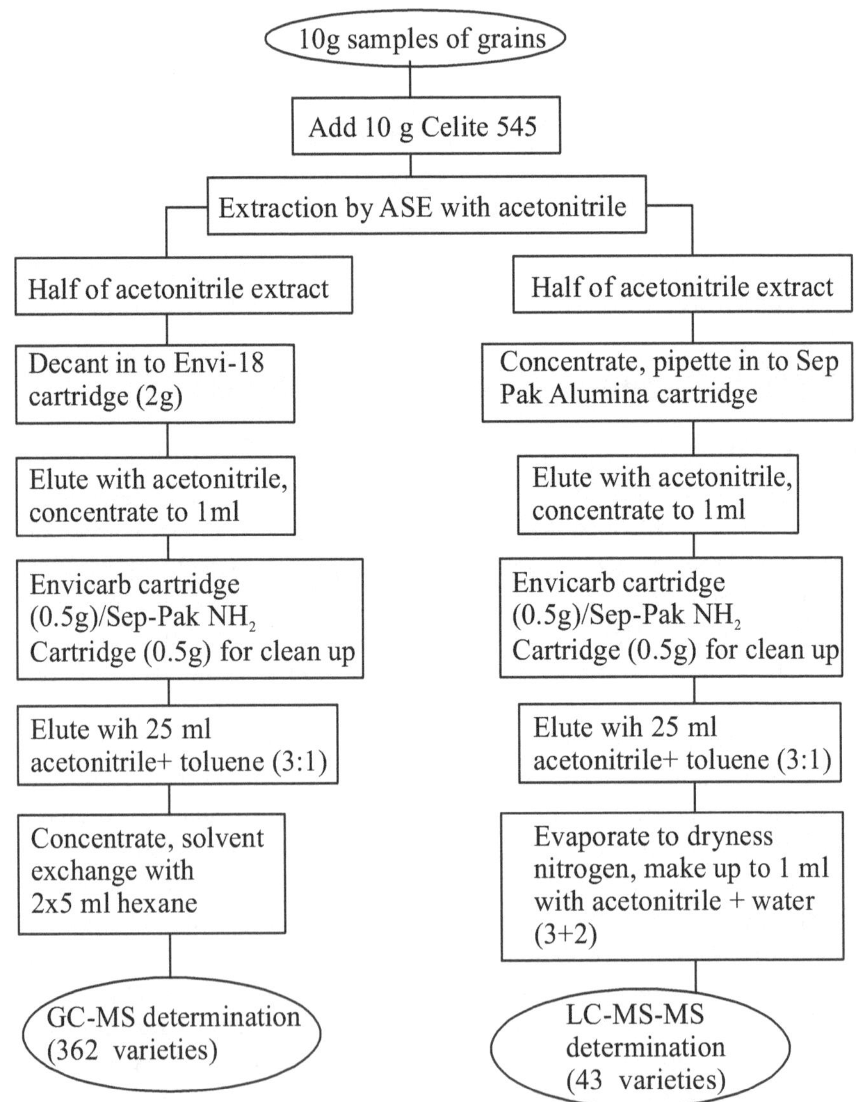

Figure 10.8: Analysis of multiresidues in food grains

10.3.2 Estimation of Multiresidues in Honey

Albero et al., (2004) developed an analytical method for the simultaneous determination of 51 pesticides in commercial honeys. Honey (10 g) was dissolved in water/methanol (70:30; 10mL) and transferred to a C_{18} column (1 g) preconditioned with acetonitrile and water. Pesticides were subsequently eluted with a hexane/ethyl acetate mixture (50:50) and determined by gas chromatography with electron impact mass spectrometric detection in the selected ion monitoring mode (GC-MS-SIM). Recovery studies performed at 0.1, 0.05, and 0.025 μg/g fortification levels for each pesticide, were >86% with relative standard deviations of <10%. Good resolution of the pesticide mixture was achieved in ~41 min with the detection limits of the method ranged from 0.1 to 6.1 μg/kg for the different pesticides studied. The developed method is linear over the range assayed, 25-200 μg/L, with determination coefficients of >0.996. The proposed method was applied to the analysis of pesticides in honey samples, and low levels of a few pesticides (dichlofluanid, ethalfluralin, and triallate) were detected in some samples.

10.3.3 Estimation of Multiresidues in Fruits and Vegetables

A simple, rapid, and sensitive multi-residue method for the determination of 205 pesticides commonly used for crop protection was developed by Xiu et al.,(2009). Pesticide residues were extracted from samples with acetonitrile and a mini-solid phase extraction filter. No additional concentration or clean-up steps were performed. The developed method was linear over the range assayed (5–50 kg L^{-1}), with determination coefficients >0.99. The limits of quantification were in the range 0.0005–0.6 mg kg^{-1} and the limits of detection were between 0.00015 and 0.2 mg kg^{-1}. This method was applied to determine pesticide levels in vegetables, fruits, and beans.

10.3.4 Estimation of Multiresidues in Honey, Fruit Juice and Wine

Pang et al., (2006) developed a multi-residue method for the determination of 450 pesticide residues in honey, fruit juice and wine using double-cartridge solid-phase extraction (SPE), gas chromatography-mass spectrometry (GC-MS) and liquid chromatography-tandem mass spectrometry (LC-MS-MS). The method development was based on an appraisal of the characteristics of GC-MS and LC-MS-MS for 654 pesticides as well as the efficiency of extraction and purification from honey, fruit juice and wine. Samples were first diluted with water plus acetone, then extracted with portions of dichloromethane. The extracts were concentrated and cleaned up with graphitized carbon black and aminopropyl cartridges stacked in tandem. Pesticides were eluted with acetonitrile + toluene, and the eluates were concentrated. For 383 pesticides, the eluate was extracted with hexane twice and internal standard solution was added prior to GC-MS determination. For 67 pesticides, extraction was with methanol prior to LC-MS-MS determination. The limit of detection for the method was between 1.0 and 300 ng g(-1) depending on each pesticide analyte. At the three fortification levels of 2.0-3000 ng g(-1), the average recovery rates were between 59 and 123%, among which 413 pesticides (92% of the 450) had recovery rates of 70-120% and 35 pesticides (8% of the 450) had recovery rates of 59-70%. There were 437 pesticides (97% of the 450) with a relative standard deviation below 25%; there were 13 varieties (3% of the 450) between 25.0 and 30.4%.

Multiresidue methods are used to carry out the analysis of environmental components (air, soil, water) or food commodities which contain residues of large number of pesticides. For extraction of pesticides, such a solvent is selected in which large number of pesticides have good solubility even if the recoveries are low in some case. Analysis is carried out on gas liquid chromatography equipped with capillary column preferably split less system and specific detectors to separate a number of pesticides and their metabolites.

Part C

10.4 Quick, Easy, Cheap, Effective, Rugged and Safe (QuEChERS) Method

It is a relatively new multi-residue method for determining pesticide residues in different matrices. The change in pesticide usage pattern over the past some years has necessitated to develop residue analytical techniques capable of qualitative detection and quantitative estimation of the multiresidues resulting from application of different xenobiotics of intra and inter class chemicals on field crops. With more public awakening and WTO era, multiresidue analysis of pesticides in fruits, vegetables and other foods has become primary function of several regulatory, industrial and contract laboratories throughout the world. According to a rough estimate > 2, 00,000 food samples are analysed world wide every year for pesticide residues to meet a variety of purposes. Once analytical quality requirements, viz. trueness, precision, sensitivity, selectivity, and analytical scope, have been met to suit the need for any particular analysis, all other purposes for analysis go in favour of practical benefits (high sample throughput, ruggedness, ease- of –use, low cost and labour, minimal solvent usage and waste generation, occupational and environmental-friendliness, small space requirements and few material and glass ware needs). A number of analytical methods designed to determine multiple pesticide residues have been developed in the time since this type of analysis became important some 2-3 decades ago. However, few if any of these methods can simultaneously achieve high quality results for a wide range of pesticides and the practical benefits desired by all laboratories. In 2003, the QuEChERS method for pesticide residue analysis was introduced which provides high quality results in a fast, easy and inexpensive approach. Follow up studies have further validated the method for >200 pesticides, improved results for the remaining few problematic analytes and tested it in fat containing matrices. This method has been explined in detail by Lehotay (1994).

10.4.1 Principle of the QuEChERS Method

The QuEChERS method, known as the quick, easy, cheap, effective, rugged and safe for pesticide residues, involves the extraction of the sample with acetonitrile (MeCN) containing 1% acetic acid (HAc) and simultaneous liquid-liquid partitioning formed by adding anhydrous magnesium sulphate ($MgSO_4$) plus sodium acetate (NaAc) followed by a simple cleanup step known as dispersive solid-phase extraction (SPE). The method is carried out by shaking a Teflon centrifuge tube which contains 1 ml of 1% of HAc in MeCN plus 0.4 g anh. $MgSO_4$ and 0.1 g anh. NaAc per g sample. The tube is then centrifuged and portion of the extract is transferred to a tube containing 50 mg primary secondary amine (PSA) sorbent to remove fatty acids among other components plus 150 mg anhydrous $MgSO_4$ per ml extract to reduce the remaining water in the extract. (the dispersive-SPE cleanup step). Then, the extract is centrifuged and transferred to autosampler vials for concurrent analysis by gas chromatography/mass spectrometery (GC/MS) and liquid chromatography/tandem mass spectrometry (LC/MS-MS). Different options are possible depending on alternate analytical instrumentation available, desired limit of quantitation (LOQ), scope of targeted pesticides and tested matrices.

The final extract concentration of the method in MeCN is 1 g/ mL. To achieve < 10 ng/g limit of quantitation (LOQ) in GC/MS, large volume injection of 8 μL is typically needed, or the final extract can be concentrated and solvent exchanged to toluene (4 g/mL) in which case 2 μL injection provides the anticipated degree of sensitivity. If MS instruments are not available in the laboratory, other options are also possible to analyse the samples using LC and GC coupled to element selective detectors.

10.4.2 Advantages of QuEChERS Method Over The Traditional Multiresidue Methods

The QuEChERS method has several advantages over most traditional methods of analysis in different ways:

(i) high recoveries (>85%) can be achieved for a wide polarity and volatility range of pesticides, including notoriously difficult analytes,

(ii) very accurate results are achieved because an internal standard (I.S.) is used to correct for commodity to commodity water content differences and volume fluctuations,

(iii) high sample throughput of about 10-20 pre-weighed samples in about 30-40 min is possible,

(iv) solvent usage and waste is very small and no chlorinated solvents are used,

(v) a single person can perform the method without much training or technical skill,

(vi) very little glassware is used,

(vi) method is quite rugged because extract cleanup is done to remove organic acids,

(vii) very little bench space is needed, thus the method can be done in a small laboratory if needed,

(viii) the MeCN is added by dispenser to an unbreakable vessel that is immediately sealed, thus solvent exposure to the worker is minimal,

(ix) the reagent costs in the method are very inexpensive and

(x) few devices are needed to carry out sample preparation.

References

Albero,B., Brunete, C.S., and José L., Tadeo, J.L., 2004. Analysis of pesticides in honey by solid-phase extraction and Gas chromatography-Mass Spectrometry. *J. Agric. Fd. Chem.* 52 (19): 5828–5835.

Kathpal,T.S., Kumari, Beena, Singh, Shashi and Singh, Jagdeep, 2004. Multiresidue analysis of bovine and human milk in cotton growing belt of Haryana. In: *Pesticide: Environment and food security.* (eds P.Dureja, D.B.Saxena,Jitender Kumar, Madhuban Gopal, Shashi Bala Singh and R.S.Tanwar) Society of Pesticide Science, India, New Delhi**:** 140-148.

Kumari, Beena, Gulati, Rachna and Kathpal, T.S., 2003. Monitoring of pesticidal contamination in honey. *The Korean J. of Apiculture.* 18 (2):155-160.

Kumari, Beena, Kumar, R. and Kathpal, T.S., 2001.An improved multiresidue procedure for determination of 30 pesticides in vegetables. *Pestic. Res. J.* 13(1): 32-35.

Kumari, Beena, Madan, V.K. and Kathpal T. S., 2006. Monitoring of pesticide residues in fruits. *Environ. Monit. and Assess.* 123: 407-412.

Kumari, Beena, Madan, V.K. and Kathpal, T. S., 2008. Status of insecticide contamination of soil and water in Haryana, India. *Environ. Monit. and Assess.* 136: 239-244.

Lehotay, S.J., 2004. Quick, easy, cheap, effective, rugged and safe (QuEChERS) Approach for determining pesticide residues. In : *Pesticide Analysis in Methods in Biotechnology* (eds. Vidal Martinez, J.L. and Garrido Frenich,A.), Humana Press, USA

Nakamura, Y., Tonogaiy, Sekiguchi, Y., Tsumura, Y., Nishida, N., Takakura, K., Isechi, M., Yuasa, K., Nakamura, M., Kifune, N., Yamamoto, K., Terasewa, S., Oshima, T., Miyata, M., Kamakura,K. and Ito, Y.: 1994, 'Multi-residue analysis of 48 pesticides in agricultural products by capillary gas chromatography', *J. Agric. Fd. Chem.* 42: 2508–2518.

Pang, F., Liu, Y.M., Fan,C.L., Zhang,J.J., Cao, Y.Z., Li, X.M., Li, Z.Y., Wu, Y.P and Guo,T.T.,2006. Simultaneous determination of 405 pesticide residues in grain by accelerated solvent extraction then gas chromatography-mass spectrometry or liquid chromatography-tandem mass spectrometry. *Anal Bioanal Chem.* 384(6):1366-408.

Pang, G.F., Fan ,C.L., Liu, Y.M., Cao, Y.Z., Zhang, J.J., Fu, B.L., Li, X.M., Li, Z.Y., Wu, Y.P.,2006. Multi-residue method for the determination of 450 pesticide residues in honey, fruit juice and wine by double-cartridge solid-phase extraction/gas chromatography-mass spectrometry and liquid chromatography-tandem mass spectrometry. *Food Addit. Contam.* 23(8): 777-810.

Rathi, Anju, Kumari, Beena, Gahlawat, S.K., Sihag, R.C. and Kathpal, T.S., 1997. Multiresidue analysis of market honey samples for pesticidal contamination. *Pestic. Res. J.* 9(2): 226-230.

❑❑❑

Chapter 11
Glossary

Glossary of terms relating to pesticides is recommended by International Union of Pure and Applied Chemistry (IUPAC), Holland (1996) and the following definitions of terms are among those, which have been adopted by the Codex Alimentarius Commission.

A

Abiotic degradation: Degradation of a pesticide via purely physical or chemical mechanisms. Examples include hydrolysis and photolysis.

Absorption: Transfer of a component from one phase to another. Movement of a pesticide from the environment (e.g. water, ingested food, leaf surface) across a biological membrane into an organism.

Acceptable daily intake (ADI): Estimate of the amount of a pesticide in food and drinking water which can be ingested daily over a lifetime by humans without appreciable health risk. It is usually expressed in milligrams per kilogram of body weight.

Or

Acceptable daily intake of a chemical is the daily intake, which during an entire lifetime, appears to have no appreciable risk to the health of the consumer on the basis of all the known facts at the time of the evaluation of the chemical. Without appreciable risk is understood to imply "as a matter of practical certainty". It is expressed in milligrams of the chemical per kilogram of body weight. ADIs are derived from the results of long term feeding studies with laboratory animals. A safety factor of 100 is applied to express the non-observed adverse effect level (NOAEL) in the most sensitive animal studied.

$$\text{ADI for humans (mg/kg body weight/day)} = \frac{\text{NOAEL animal studies (mg/kg body weight/day)}}{\text{100 (safety factor)}}$$

Action level (regulatory): (i) For food commodities, an administrative maximum residue limit (MRL) used by regulatory authorities to initiate action where no legally defined MRL has been established. (ii) For the environment concentration of a pesticide in air, soil or water at which emergency measures or preventative actions are to be taken.

Action limits (analytical quality control): Limits for measurements on reference material or spiked samples which indicate when an analytical procedure is not performing adequately and requires immediate action before data can be reported.

Active ingredient (a.i.): Pesticide present in a formulation as described by the common name. The part of a pesticide formulation from which the biological effect is obtained.

Accuracy (of measurement): Closeness of agreement between the result of a measurement and the (conventional) true value of the measure and.

Note 1. Use of the term precision for accuracy should be avoided.

Note 2. True value is an ideal concept and, in general, cannot be known exactly.

Acute toxicity: Ability of a substance to cause adverse effects within a short period following dosing or exposure.

Adjuvant: Formulant designed to enhance the activity or other properties of a pesticide mixture.

Adsorption: Enrichment of one or more components in an interfacial layer.

Adverse effect: Change in morphology, physiology, growth development or lifespan of an organism which results in impairment of functional capacity or which increases susceptibility to the harmful effects of other environmental influences.

Aerobic: Conditions under which molecular oxygen serves as the terminal electron acceptor in respiration or in metabolic oxygenation.

Aerosol: System of fine solid or liquid particles (<30 μm diameter) dispersed in a gas. Aerosol cans using an inert compressed propellant are a common means of dispensing insecticides for domestic use.

AFID: Alkali flame-ionisation detector or detection for gas chromatography.

Aged residue: Pesticide and degradates present in an environmental system after application and following a period long enough to allow transport, adsorption, metabolism and dissipation processes to alter the distribution and chemical nature of some of the applied pesticide.

Aggregate sample: Sample made up of set proportions of other samples, typically an average by weight.

Aglycon: Non-sugar part of a glycoside or glucuronide conjugate derived from the pesticide.

Agrochemical: Agricultural chemical used in crop and food production including pesticide, feed additive, veterinary drug and related compounds.

Aliquot: Known fractional portion of a homogeneous material. The term is usually applied to volumetric sub-sampling of fluids.

Anaerobic: Condition under which reductive conditions prevail.

Analysis: Analysis is a study or examination of something, in detail, to discover more about it.

Analytical range: Measurement range of a test method where the performance has been validated and quality standards such as action limits have been developed.

Analytical standard (pesticide): Pesticide reference material of high and defined purity (generally >95%) for preparation of calibration standards.

Animal feed: Animal feed means harvested fodder crops, by-products of agricultural crops and other products of plant or animal origin which are used for animal feeding and which are not intended for human consumption.

B

Bait: A food or other substance used to attract a pest to a pesticide or trap where it can be destroyed.

Batch: Quantity of material which is known or assumed to be produced under uniform conditions.

Benthos: Non-planktonic animals (not being suspended in water) associated with freshwater substrata (upper layer of the sediment in rivers and ponds) at the sediment-water interface.

Bioaccumulation: Progressive increase in the amount of a substance in an organism or part of an organism which occurs because the rate of intake exceeds the organism's ability to remove the pesticide from the body.

Bioactivation: Transformation of a pesticide within an organism into a more biochemically active metabolite.

Bioconcentration: Process leading to a higher concentration of a pesticide in an organism than in environmental media to which it is exposed.

Bioconcentration factor (BCF): Ratio between the concentration of pesticide in an organism or tissue and the concentration in the environmental matrix (usually water) at apparent equilibrium during the uptake phase.

Bioavailability: Extent to which a pesticide residue can be taken up into an organism from its food and environment, and the rate at which this occurs.

Biodegradation: Conversion or breakdown of the chemical structure of a pesticide catalysed by enzymes in *vitro* or in *vivo,* resulting in loss of biological activity. For hazard assessment, categories of chemical degradation include:

1. Primary- loss of specific activity
2. Environmentally acceptable- loss of any undesirable activity (including any toxic metabolites)
3. Ultimate- mineralization to small molecules such as water and carbon dioxide

Biological assessment of exposure: Assessment of exposure of a living organism to pesticides using biological specimens (blood, urine etc.) taken in the environment (workplace, field etc.) with analysis either directly by chemical determination of parent or metabolite, or indirectly by measurement of a relevant biochemical parameter (e.g. plasma cholinesterase activity for organophosphorus compounds).

Biomagnification: Bioaccumulation of a pesticide through an ecological food chain by transfer of residues from the diet into body tissues. The tissue concentration increases at each trophic level in the food web when there is efficient uptake and slow elimination.

Biomarker: Indicator (molecular, biochemical, cellular or organism) signalling an event or condition in a biological system or sample and giving a measure of exposure to, effect of, or susceptibility to, a xenobiotic.

Biomass: The total living mass in a defined segment of an ecosystem expressed as the living weight per unit area or mass. Soil microbial biomass is often used as an indication of potential microbial activity level in soil.

Biometer flask: Experimental apparatus commonly used in laboratory studies of pesticide degradation in soil. Contains separate compartments for aerobic incubation of soil and for media to trap carbon dioxide and volatile products.

Biopesticide: Pesticide of biological origin including microorganisms e.g. *Bacillus thuringiensis* and natural products e.g. rotenone, pyrethrins.

Biotransformation: Conversion of the chemical structure of a pesticide catalysed by enzymes in *vitro* or in *vivo.*

Biotransformation pathway: Sequence of the changes, occurring in the structure of a pesticide when it is introduced into a specific biological test system.

Blank material (sample): Laboratory simulated test material known to be free of the pesticide being analysed. A portion of blank material is used to test the method, apparatus and reagents for interferences or contamination.

Bound residue: Chemical species in soil, plant or animal tissue originating from a pesticide, (generally radio labeled) that are unextracted by a standard method, such as Soxhlet solvent extraction, which does not significantly change the chemical nature of the residues. These unextractable residues are considered to exclude small fragments recycled through metabolic pathways into natural products.

Buffer zone: Distance for environmental protection between the edge of an area where pesticide application is permitted and a sensitive non-target area e.g. water course.

C

Carrier: Solid formulant added to a technical material as an absorbent or diluent.

Carryover (analytical): Unintended contamination of a sample undergoing analysis with material from a previous sample.

Carryover (field): Persistence of pesticide residues in soil after use in one crop such that uptake is observed in a succeeding, possibly more sensitive, crop.

Catabolism: Oxidative biodegradation of a pesticide to provide chemically available energy and generate metabolic intermediates.

Catchment: Land and water confined within a single drainage basin.

Certified reference material: Reference material, accompanied by a certificate, whose pesticide concentrations are certified by procedures which establish their traceability and for which each certified concentration is accompanied by an uncertainty at a stated level of confidence. Storage conditions and period for which the certification remains valid may also be included for unstable materials.

Chronic effect: Consequence of exposure to a pesticide which arises slowly and has a long-lasting, often irreversible, course.

Chronic exposure: Continued exposures occurring over an extended period of time, or a significant fraction of the lifetime of the exposed individuals or test species.

Chronic toxicity: Capacity for a pesticide to produce injury following chronic exposure or to produce effects which persist whether or not they occur immediately upon exposure or are delayed.

Co-metabolism: Microbial metabolism of a pesticide where the derived energy is not used to support microbial growth.

Common moiety: Molecular sub-unit which is common to the structures of several pesticides or metabolites.

Community: Assembly of populations of different species of living organisms (quite often interdependent on and interacting with each other) within a specified location in space and time.

Compartment: Part of an organism or ecosystem considered as an independent system for purposes of assessment of uptake, distribution and dissipation of a pesticide.

Compliance (residue): Meeting of official maximum residue limit (MRL) standards by residue levels in food consignments sampled and tested by approved methods.

Composite sample: Combined increment samples, or combined replicate samples, or combined samples from replicate trials. Preferred term to bulk sample which is ambiguous.

Concentration-effect relationship: Association between the exposure concentration of the pesticide and the magnitude of the resultant continuously graded change either in an individual organism or in a population.

Conjugation: Biosynthetic reaction in which a pesticide or its metabolite is linked to an endogenous compound.

Contaminant: 1. Minor impurity in a substance.

2. Extraneous material added to a sample prior to or during chemical or biological analysis.

3. Unintended pesticide residue in an agricultural commodity or environmental compartment (e.g. ground water).

Control sample (field): Sample from a field test plot to which no pesticide was applied (a zero rate sample) or which received chemical treatments identical to the test plots except for the test chemical.

Critical concentration: It is the lowest concentration of a pesticide in an environmental compartment at which adverse effects on organisms are likely to be observed (95% probability).

Critical load: Amount of a pesticide leading to a critical concentration when received by an environmental compartment.

Cumulative effect: Overall adverse change which occurs when repeated doses of a pesticide have biological consequences which are additive.

Cut-off value: Numerical value set by regulatory authorities representing the limit of acceptability for a property or behaviour of a compound for the final step in tiered assessment schemes.

D

Degradate: Chemical product resulting from degradation of a pesticide.

Degradation: Process by which a pesticide is broken down to simpler structures through biological or abiotic mechanisms. Synonyms include breakdown and decomposition.

Deposit: The amount of initially laid down pesticide on the surface is called deposit.

Desorption: Depletion of one or more components in an interfacial layer.

Detoxification: Processes of chemical modification which makes a pesticide less toxic.

Diluent: Liquid or solid material used to dilute a concentrated pesticide formulation prior to application. Most commonly water for spray application.

Dislodgeable residue: Portion of a pesticide residue on treated vegetation that is readily removable and may be used as an index for risk to farm workers. Generally measured by the residue removed when leaf discs are shaken briefly in water.

Dissipation: Loss of pesticide residues from an environmental compartment due to degradation and transfer to another environmental compartment.

Dissipation and persistence: In nature disappearance of residues takes place in 2 steps. The first step is the initial phase in which the disappearance of the residue is fast. This phase is called "Dissipation". The second phase, in which there is a slow decrease in the amount of residue, is known as "Persistence". The main difference between these is that the dissipation follows the law of "first-order kinetics", whereas the persistence does not follow this law because of the storage of translocated insecticides and degradation of various rates.

Dissipation time 50% (DT_{50}): Time required for one-half the initial quantity or concentration of a pesticide to dissipate from a system. No assumption as to the rate equation is made.

Dose-effect relationship: Graded relationship between the dose of the pesticide to which the organism is exposed and the magnitude of a defined biological effect either in an individual organism or in a population.

Dose-response relationship: Association between dose and the incidence of a defined biological effect in an exposed population.

Drift control agent: Formulant that control the distribution of spray droplet sizes and prevents production of excessive fines.

Dry weight basis: Pesticide residue concentration reported as if the residue were wholly contained in the dry matter of the sample, i.e. analytical results are corrected for the water content of the test sample. Residues in soils and feeds, and maximum residue limits (MRLs) for feedstuffs are expressed on a dry weight basis.

Dustable powder (DP): Free flowing powder suitable for dusting.

E

ECD: Electron capture detector, used in gas chromatography.

Ecosystem: Assembly of populations of different species (often interdependent on and interacting with each other) interacting with their surroundings within a specified physical location and forming a functional entity.

Ecotoxicologically (environmentally) relevant concentration (ERC): Concentration of a pesticide (active ingredient, formulations and relevant metabolites) that is likely to affect a determinable ecological characteristic of an exposed system. It is related to the toxicity characteristics, generally the no observable effect concentration, to the most sensitive species or groups of species.

Emulsifiable concentrate (EC): Liquid homogeneous formulation of a pesticide with emulsifiers in an organic solvent which forms dispersion when added to water as a diluent.

Emulsifier: Surfactant used to facilitate the preparation of a colloidal dispersion of one liquid in another liquid with which it is not miscible.

Endocon: That portion of a conjugated metabolite which is derived from natural products of the metabolizing organism such as sugars and organic acids.

Endpoint: Measurable ecological or toxicological characteristic or parameter of the test system (usually an organism) that is chosen as the most relevant assessment criterion (e.g. death in an acute test or tumour incidence in a chronic study).

Enhanced degradation: Increased rate of degradation of a pesticide in soil or other environmental matrix by a population of microorganisms that has adapted to metabolise it through previous exposure to it or a similar chemical. Synonyms include accelerated degradation and enhanced biodegradation.

Enterohepatic circulation: Cyclical process in which a pesticide residue is absorbed and transported to the liver, metabolised (often including conjugation), transported to the intestine by the bile, reabsorbed (often after deconjugation), and transported to the liver for further metabolism.

Environmental impact assessment: Assessment of the potential releases of a pesticide to the environment and their potential effects upon the environment and its components including man.

Environmental risk: Probability that an adverse effect on humans or the environment will be observed for a given exposure to a pesticide based on the frequency of occurrence and the sensitivity of the system.

Estimated daily intake (EDI): Prediction of the daily intake of a pesticide residue, based on the most realistic estimation of residues in food items and the best available food consumption data for a specific population. Residue levels are estimated taking into account known uses of the pesticide, the proportion of commodity treated and the quantity of contaminated commodities. The EDI is expressed in milligrams of residues per person.

Estimated environmental concentration (EEC): Predicted concentration of a pesticide within an environmental compartment based on estimates of quantities released, discharge patterns and inherent disposition of the pesticide (fate and distribution) as well as the nature of the specific receiving ecosystems.

Estimated maximum daily intake (EMDI): Prediction of the maximum daily intake of a pesticide residue, based on the assumptions of average daily food consumption per person and maximum residues in the edible portion of a commodity, corrected for the reduction or increase in residues resulting from preparation, cooking, or commercial processing. The EMDI is expressed in milligrams of residues per person.

Exocon: That portion of a conjugated metabolite which is derived from the parent pesticide.

Exposure: Concentration or amount of a pesticide that reaches the target population, organism, tissue or cell, usually expressed in numerical terms of concentration, duration and frequency. Also the process by which a substance becomes available for absorption by the target population, organism, tissue or cell, by any route.

Exposure assessment: Process of measuring or estimating concentration, duration and frequency of exposures to pesticide present in environment or, if estimating hypothetical exposures, that might arise from the release of the pesticide into the environment.

Extractability: Degree to which a pesticide residue may be removed from a matrix (e.g. soil) through use of appropriate extraction techniques.

Extraneous residue limit (ERL): The ERL refers to a pesticide residue or a contaminant arising from environmental sources (including agricultural uses) other than the use of a pesticide or contaminant substance directly or indirectly on the commodity that is recommended to be permitted in or on a feed or food commodity. It is the maximum concentration of a pesticide residue or contaminant that is recommended by *Codex Alimentarius Commission* to be legally permitted or recognized as acceptable in or on a feed, agricultural commodity or animal feed. The concentration is expressed in milligrams of pesticide residues or contaminant per kilogram of the commodity.

F

Fat basis: Residues and maximum residue limits (MRLs) of fat-soluble pesticides in animal commodities may be expressed in terms of their concentration in the fat rather than the whole product.

FID: Flame ionisation detector for gas chromatography or HPLC.

Field drainage: Removal of excess water from soil and transport to surface waters in order to improve soil productivity and trafficability.

First-order kinetics: The rate of disappearance is related to the amount deposited (the rate of reaction is directly proportional to the concentration of reactants). Thus a plot of the log amount of pesticide left at the initial deposit site against time (usually days) should be a straight line.

Food chain - primary producers: Autotrophic organisms (e.g. algae, higher plants) which convert inorganic compounds during the process of photosynthesis or chemosynthesis into organic compounds (cell material) of higher energy content. These organisms represent the first trophic level of the food chain.

Food chain - secondary producers: Heterotrophic organisms (e.g. animals) using organic substances as a carbon and energy source.

Food chain - primary consumers: Heterotrophic organisms (e.g. filter feeding invertebrates such as daphnia species) using organic substances directly from primary producers (e.g. algae) as a carbon and energy source.

Food chain- secondary consumers: Heterotrophic organisms (e.g. predator animals) feeding on primary consumers.

Food chain - primary decomposers: Heterotrophic organisms (e.g. bacteria) using dead organic matter from all trophic levels as a carbon and energy source.

Food chain - secondary decomposers: Heterotrophic organisms (e.g. certain soil fungi, Collembola, worms) using already partially decomposed organic matter as a carbon and energy source.

Food-factor: It is the average fraction of total diet made by the food or class of food under question. For example, if the total food consumed by a person per day is 2 kg and the treated food (containing 1 ppm pesticide) be 500g, the food factor will be 500/2000=0.25. It is used in determining the daily intake of pesticide, which is used for calculating MRL.

Formulant: Any added material in a pesticide formulation other than the biologically active ingredient(s). This may include carrier or other substances which enhance the biological activity or physio-chemical properties of the formulation.

FPD: Flame photometric detector for gas chromatography.

Fresh weight basis: Pesticide residues are reported on the laboratory sample as it is received, with no allowance for the moisture content. Maximum residue limits (MRLs) and pesticide residues in food commodities are expressed in this way.

Freundlich isotherm: Empirical relationship describing the adsorption of a solute from a liquid or gaseous phase to a solid in which the quantity of material adsorbed per unit mass of adsorbent is expressed as a function of the equilibrium concentration of the sorbate.

Fumigation: Use of a pesticide in gas or vapour form.

FTIR: Fourier transform infrared spectroscopy.

G

GC-EC: Gas chromatography with electron capture detector.

GC-MS: Gas chromatography-mass spectrometry.

GC-MSD: Gas chromatography with mass-selective detection (usually low resolution mass spectrometry using selected ions).

GLC: Gas liquid chromatography

GLP study plan: The document which determines the entire scope of a study conducted under GLP. A written study plan must be completed and approved by the Study Director before a study starts. It contains information such as the title of study; name or code of test and reference substances; name and address of sponsor, test facility, study director and principal investigators; dates for start and end of study; methods including relevant standard operating procedures (SOPs); list of material to be archived.

Good agricultural practice (GAP): The GAP includes the nationally authorised safe uses of pesticides under actual conditions necessary for effective and reliable pest control. It encompasses a range of levels of pesticide applications up to the highest authorised use, applied in a manner that leaves a residue, which is the smallest amount practicable. Authorised safe uses include nationally registered or recommended uses that take into account public and occupational health and environmental safety considerations. Actual conditions include any stage in the production, storage, transport, distribution and processing of food commodities and animal feed.

GLP certificate: Certificate of test facility compliance with a national GLP programs.

Good laboratory practice (GLP): The formalized process and conditions under which laboratory studies on pesticides are planned, performed, monitored, recorded, reported and audited. Studies performed under GLP are based on the national regulations of a country and are designed to assure the reliability and integrity of the studies and associated data. The US-EPA GLP definition also covers field experiments.

GLP quality assurance unit (QAU): Sub-section of the test facility, separate from actual testing, responsible for internal audits of the facility and its study reports to ensure compliance with GLP. The QAU is also responsible for the administration and training in all aspects of the quality assurance unit.

GLP standard operating procedure (SOP): Written procedure, authorized by management, which describes how to perform a certain routine test, or activity not specified in detail in study plans or test guidelines.

GPC: Gel permeation chromatography.

Granule: Solid formulation comprising particles of defined size >80 μm diameter, for application without further dilution, usually to soil.

Ground water: Water present in the saturated subsurface zone of the soil profile, where all open spaces/pores in the sediment and rock are filled with water.

Guideline level: Maximum concentration of a pesticide residue in or on a feed or food commodity, resulting from a use reflecting good agricultural practice, but where an acceptable daily intake has not been estimated.

Guideline level (GL): It is used to assist authorities in determining the maximum concentration of a pesticide residue resulting from a use reflecting good agricultural practice in situations where an acceptable daily intake or temporary acceptable daily intake for the pesticide has not been estimated. The concentration is expressed in milligrams of pesticide residue per kilogram of the commodity.

Guideline value (GV): Maximum recommended pesticide residue in an environmental medium that ensures aesthetically pleasing air, water or food and does not constitute a significant risk to the user.

H

Half-life ($t_{1/2}$): Time taken for the concentration of a pesticide in a compartment to decline by one half. Usually an estimate based on observed dissipation over several half-lives that can be described by first order kinetics with rate constant k, $t_{1/2} = 0.693/k$.

Or

Half-life: It is the time in which half the amount of initial deposit is eliminated by reacting or otherwise getting dissipated. Half-life values or 90 -100% disappearance values are used to express the approximate rates of residue disappearance. These values are largely determined by the initial fast phase of dissipation through which the major portion of the initial deposit is often eliminated. Such values can serve as a general quick reference to the rate of insecticide disappearance.

Hazard: Set of inherent properties of a pesticide which gives potential for adverse effect to man or the environment under conditions of its production, use or disposal, and depending on the degree of exposure.

Hazard assessment: Determination of factors controlling the likely effects of a hazard such as mechanism of toxicity, dose-effect relationships and worst case exposure levels. This is the prelude to risk assessment.

Hazardous distance for the most sensitive effect (HDSE): Statistically determined safety margin corresponding to a distance from treated areas at which protection of the terrestrial environment can be adequately achieved as measured by the most sensitive non-target species.

Health advisory level (HAL): Estimate of upper concentration limit for a pesticide in drinking water that can be consumed for a lifetime without adverse effects. HALs generally does not have formal legal significance but have been used, particularly in the USA, for preliminary risk assessment.

HPLC: High performance liquid chromatography.

HPTLC: High performance thin layer chromatography.

HRGC: High resolution gas chromatography (GLC with narrow bore capillary columns).

Hydrolysis: Reaction in which a chemical bond is cleaved and a new bond formed with the oxygen atom of a molecule of water.

I

Identification: Process of unambiguously determining the chemical identity of a pesticide or metabolite in experimental or analytical situations.

Immobilization:
1. Process leading to restricted mobility of a pesticide in plant or soil due to strong binding.
2. Incorporation of terminal pesticide degradates into complex organic forms in microbial or plant tissue.

Immunoassay: Ligand-binding assay based on antibodies capable of specific binding to the pesticide analyte. Most commonly used in a competitive binding format where analyte molecules compete with a specific antigen complex labelled for detection using a radioisotope (radio immunoassay - RIA) or enzyme (enzyme linked immunoassay - ELISA).

Impurity: By-product of the manufacture or storage of a pesticide. Impurities require definition, evaluation and regulation (if toxicologically significant).

Increment sample: An individual portion (unit) of material collected by a single operation of a sampling device from bulk materials or large units.

Incurred residue: Residues in a commodity resulting from specific use of a pesticide, consumption by an animal or environmental contamination in the field, as opposed to residues from laboratory fortification of samples.

Inert ingredient: Formulant which by itself does not add materially to effectiveness for the purpose for which the preparation is intended e.g. solvent, emulsifier, diluent, and carrier.

In-life phase: Phase of a study following treatment in which the test system is alive/growing.

Intake study: It is a study designed to measure or estimate actual dietary exposures of consumers to pesticide residues or contaminants in order to compare such exposures to the acceptable daily intakes for pesticides or contaminants.

***In vitro*:** 'In glass', referring to use of a cell line, microorganisms or biochemically active fraction derived from an organism in laboratory studies of biological activity, or metabolism or toxicity.

***In vivo*:** Use of the whole living organism in studies of biological metabolism or toxicity.

L

Laboratory sample: Sample or subsample(s) sent to or received by the laboratory.

Lag phase: Period which may precede commencement of rapid degradation of a pesticide by a microbial population. It is the period needed either for induction of microbial enzymes or for growth of the microbial population to adequate size.

Leaching: Process by which a pesticide moves downward through the soil profile in the aqueous phase.

Leachate: Aqueous phase percolating through a soil profile or a soil column.

Limit of detection (LOD): Lowest concentration of a pesticide residue in a defined matrix where positive identification can be achieved using a specified method.

Limit of determination (LODe): The LODe is the lowest concentration of a pesticide residue or contaminant that can be identified and quantitatively measured in a specified food, agricultural commodity, or animal feed with an acceptable degree of certainty by a regulatory method of analysis.

Limit of quantitation (LOQ): Lowest concentration of a pesticide residue in a defined matrix where positive identification and quantitative measurement can be achieved using a specified method. The term limit of quantitation is preferred to limit of determination to differentiate it from LOD. LOQ has been defined as 3 times the LOD or as 50% above the lowest fortification level used to validate the method.

Limit of reporting: Practical limit of residue quantitation at or above the LOQ. The limit of quantitation for a defined matrix and method may vary between laboratories or within the one laboratory from time to time because of different equipment, techniques and reagents.

Lipophilicity: Affinity for fat as described by partitioning behaviour between water and an immiscible organic solvent, favouring the latter, and which correlates with bioconcentration.

Lot: Quantity of material which is assumed to be a single population for sampling purposes.

M

Market basket survey: Pesticide residue monitoring on a wide range of food items collected from consumer points of sale and in proportions approximating consumption patterns in the local population. Samples are prepared for analysis according to Codex guidelines i.e. minimal preparation.

Matrix: The material or component sampled for pesticide residue studies.

Maximum residue limit (MRL): Maximum concentration of a residue that is legally permitted or recognised as acceptable in, or on, a food, agricultural commodity or animal feedstuff as set by Codex or a national regulatory authority. The term tolerance used in some countries is, in most instances, synonymous with MRL. Normally expressed as mg/kg fresh weight.

Maximum permissible intake (MPI): The maximum permissible intake (MPI) is derived by multiplying ADI with the average body weight.

Maximum tolerated dose (MTD): Highest dose of a pesticide in chronic toxicity testing that is expected, on the basis of sub-chronic studies, to produce only limited toxicity when administered for the duration of the test period.

Median effective concentration (EC_{50}): Statistically derived concentration of a pesticide in an environmental medium expected to produce a certain effect in 50% of the test organisms in a given population under defined conditions.

Median lethal concentration (LC_{50}): Statistically derived concentration of a pesticide in an environmental medium expected to kill 50% of test organisms in a given population under defined conditions.

Median lethal dose (LD_{50}): Statistically derived dose of a pesticide expected to kill 50% of test organisms in a given population under a defined set of conditions. Normally expressed as mg of test material per kg of body weight of the organism.

Metabolism: Sum total of all physical and chemical processes that take place within an organism; in a narrower sense, the physical and chemical changes that occur for a pesticide within an organism. It includes uptake and distribution within the body, changes (biodegradation), and elimination of pesticides and their metabolites.

Metabolites: Any intermediate or product resulting from metabolism.

Mineralisation: Conversion of an element from an organic form to an inorganic form. Mineralisation of pesticides most commonly refers to the microbial degradation to carbon dioxide as a terminal metabolite.

Model: Experimental or mathematical simulation of chemical behaviour in a specific environment.

Model calibration: Testing of a model with known input and output information for adjustment or estimation of factors for which data are not available.

Model, computer: Assembly of numerical techniques (algorithms), book keeping, and control language (i.e. the computer program) comprising a mathematical model and which carries out acceptance of input data and instructions through to delivery of output.

Model ecosystem: Man-made study system containing associated organism and abiotic components that is large enough to be representative of a natural ecosystem, yet small enough to be experimentally manipulated. There is some subjective differentiation between larger, outdoor model ecosystems (mesocosms) and smaller, generally indoor model ecosystems (microcosms).

Multiresidue method: Analytical method which measures a number of pesticide residues simultaneously.

N

Nebulisation: Formation of an aerosol of very small liquid particles (fog) or solid particles (smoke) from a pesticide formulation, generally for fumigation of an enclosed space such as a glass-house.

NMR: Nuclear magnetic resonance spectroscopy.

No-effect levels: After considering all available data on each compound, "No-effect" levels are determined for human safety by regulatory agencies. To regulate pesticide residues to safe levels, Joint FAO\WHO Expert Committee has introduced a concept on food additives through *Codex Alimentarius Commission.*

Non-target organism: Organism affected by a pesticide although not an intended object of its use.

No-observable effect concentration/level (NOEC/NOEL): Highest concentration or amount of pesticide in the test system that causes no observable biological effect to the target organism.

No-observable effect concentration or level (NOEC/NOEL): Highest concentration or amount of pesticide in the test system that causes no biological effect to the target organism.

Non-observed adverse effect level (NOAEL): It is the highest dose of substance that does not cause any detectable toxic effects in experimental animal studied. It is expressed in milligrams per kilogram of body weight per day.

NPD: Nitrogen-phosphorus detector or detection for gas chromatography.

O

OC: Organochlorine pesticide. Generic term for pesticides containing chlorine but commonly used to refer to older persistent materials including aldrin. BHC, chlordane, DDT, dieldrin, heptachlor, lindane and toxaphene.

Octanol/water partition coefficient (POW): Partition coefficient for a pesticide in the two-phase system octan-l-ol/water. The Pow is a distribution coefficient reflecting the relative lipophilicity of a pesticide and its potential for bioconcentration. For convenience, the value is often expressed in logarithmic (base 10) form (log Pow).

OP: Organophosphorus pesticide. Generic term for pesticides containing phosphorus but commonly used to refer to insecticides consisting of cholinesterase inhibiting esters of phosphate or thiophosphate.

P

Partition coefficient: Ratio of the concentrations of a substance in solution in two phases which are in equilibrium.

Parts per billion (ppb): Ratio of amounts expressed as parts pesticide per 10^9 sample. Strictly the quantities should be the same i.e. weight to weight (solids) or volume to volume (liquids or gases) e.g. 1 ppb = 1 µg/kg. A common usage is for weight to volume but to avoid confusion it is recommended that the SI units are used rather than ppb; e.g. µg/L.

Parts per million (ppm): Ratio of amounts expressed as parts pesticide per 10^6 sample e.g. 1 ppm = 1 mg/kg. As with ppb it is recommended that SI units are used rather than ppm, particularly for weights to volume.

PED: Plasma emission detector.

Pellet: Solid formulation of pesticide, larger than granule, often used for molluscicide formulations.

Penetrated residue: The surface residue becomes penetrated residue by migration into the sub-strata.

Persistence: Residence time of a chemical species (pesticide and/or metabolites) subjected to degradation or physical removal in a soil, crop, animal or other, defined environmental compartment.

Pest: Organism that attacks food and other materials essential to mankind or otherwise affect human beings adversely.

Pesticide: Substance or mixture of substances intended for preventing, destroying or controlling any pest, including vectors of human or animal disease, unwanted species of plants or animals causing harm or otherwise interfering with the production, processing, storage, transport, or marketing of food, agricultural commodities, wood, wood products or animal feedstuffs, or which may be administered to animals for the control of insects, mites/ spider mites or other pests in or on their bodies. The term includes substances intended for use as a plant growth regulator, defoliant, desiccant or agent for thinning fruit or preventing the premature fall of fruit, and substances applied to crops either before or after harvest to protect the commodity from deterioration during storage or transport.

Or

Pesticide means any substance intended for preventing, destroying, attracting, repelling, or controlling any pest including unwanted species of plants or animals during the production, storage, transport, distribution, and processing of food, agricultural commodities, or animal feeds or which may be administered to animals for the control of ectoparasites. The term includes substances intended for use as a plant growth regulator, defoliant, desiccant, fruit-thinning agent, or sprouting inhibitor and substances applied to crops either before or after harvest to protect the commodity from deterioration during storage and transport. The term normally excludes fertilizers, plant and animal nutrients, food additives and animal drugs.

Note: "Agricultural commodities"-include commodities such as raw cereals, sugar beet, and cottonseed that might not, in the general sense, be considered food.

Pesticide chemical name: Scientific name of a pesticide following the recommendations of IUPAC for naming of chemical compounds or other accepted naming convention (e.g. Chemical Abstracts).

Pesticide common name: Simple name assigned to a pesticide active ingredient by the International Organisation for Standardisation (ISO) to be used as a generic or non-proprietary name.

Pesticide formulation: Pesticide product offered for sale. It generally comprises active ingredient (s) adjuvant(s) and other formulants combined to render the product useful and effective for the purpose claimed.

Pesticide residue: Substance(s) which remains in or on a feed or food commodity, soil, air or water following use of a pesticide. For regulatory purposes it includes the parent compound and any specified derivatives such as degradation and conversion products, metabolites and impurities considered to be of toxicological significance.

Or

Pesticide residue means any specified substances in food, agricultural commodities, or "animal feed" resulting from the use of a pesticide. The term includes any derivatives of a pesticide, such as conversion products, metabolites, reaction products, and impurities considered to be of toxicological significance.

Note: The term "pesticide residue" includes residues from unknown or unavoidable sources (e.g., environmental), as well as known uses of the chemical.

Pesticide residue definition: The pesticide, its metabolites, derivatives and related compounds to which the maximum residue limit (MRL) applies, as specified by Codex or a national regulatory authority.

Pesticide residue enforcement: Pesticide residue monitoring program where the intention is regulatory action against non-complying consignments.

Pesticide trade name: Proprietary name assigned to a pesticide or its formulations by the company manufacturing or selling it.

Phase I metabolism: Initial biotransformation of a pesticide. These are mainly oxidative, reductive and hydrolytic processes.

Phase II metabolism: Biotransformation where the pesticide or phase I metabolite is conjugated with a naturally occurring compound (e.g. sugars, glutathione).

Phloem: Part of the plant's vascular system adapted to the transport of photosynthetic products from leaves to the rest of the plant.

Photolysis: Chemical reaction caused by light in which a bond is cleaved.

Plant growth regulator (PGR): Naturally occurring or synthetic substance which influences plant development or reproduction but has no nutritive value.

Plant protection agent: Pesticide product intended for use in agriculture to protect crops.

Pollutant: Undesirable substance introduced into a solid, liquid or gaseous environmental medium totally or partially by human activities.

Population: Assemblage of individual organisms of defined ages and growth stages belonging to one species within a specified location in space and time.

Post-emergence: Period after a crop or pest has appeared. Herbicide usage can be referred to as post-emergence (weeds) or post-emergence (crop).

Precision: Closeness of agreement between independent test results obtained under prescribed conditions.

Pre-emergence: Period before a specified crop or pest has emerged. Generally applied to timing of herbicide applications.

Preferential flow: Leaching phenomenon whereby water and a dissolved pesticide percolating down through the soil profile move more rapidly through soil macropores or sand/gravel lens than through the network of smaller pores in the bulk soil.

Pre-harvest interval (PHI): The time interval between the last application of a pesticide to a crop and harvest.

Primary sample: Collection of one or more increments or units initially taken from a population.

Note: portions may be combined (composited or aggregated sample) or kept separate.

Processed food: Product resulting from the application of physical, chemical or biological processes, or combinations of these (e.g. canning), to a primary food commodity, and intended for sale to the consumer, for use as an ingredient in the manufacture or a food product or for further processing.

Provisional tolerable daily intake (PTDI): The PTDI is a value calculated based on toxicological data. It represents tolerable human intake of an agricultural pesticide that may occur as a contaminant in food, drinking water and the environment.

Q

Quantitative structure-activity relationship (QSAR): Quantitative association between the physio-chemical properties of a pesticide or the properties of its molecular substructures and its biological properties including its non-target toxicity.

R

Random sample: Sub-set of a sampling population that is arrived at by selecting units such that each possible unit has a fixed and determinate probability of selection.

Raw agricultural commodity: Part of a crop used as a food or feed commodity directly from the harvested crop without processing.

Raw data: All original laboratory records and documentation, or verified copies thereof, including data directly entered in a computer. They are the results from the original activities and observations in a GLP study.

Recovery, analytical: Fraction or percentage of a pesticide residue recoverable following extraction and analysis of a matrix containing the pesticide.

Redox potential: Electrical potential indicating the relative activity of oxidised and reduced species. The redox potential of an environmental matrix is a measure of the extent to which oxidising species are present to act as terminal electron acceptors in respiration.

Re-entry interval: Minimum time between pesticide application and human re-entry to a treated area. Established by a regulatory authority to assure safety of workers from exposure to residues.

Reference dose: Expected dose resulting from human exposure to a pesticide at the level at which it is regulated in the environment.

Reference material: Material or substance containing pesticide of interest at levels sufficiently homogenous and well characterised to be used for the calibration of an apparatus or assessment of analytical method performance.

Registration: The process whereby the responsible national government authority approves the sale and use of a pesticide following the evaluation of scientific data demonstrating that the pesticide is effective for the purposes intended and not unduly hazardous to human or animal health or the environment.

Regulatory method: Validated analytical method which can be applied using commonly available laboratory equipment and instrumentation. A regulatory method has the precision, specificity, limit of determination, etc, needed to test compliance with the regulations.

Repeatability: For an analytical method, the closeness of agreement between results of measurements on identical test material subject to the following conditions: same analyst, same instrumentation, same location, same conditions of use, repetition over a short period of time.

Reproducibility: For an analytical method, the closeness of agreement between results of measurements on identical test material where individual measurements are carried under changing conditions such as: analyst, instrumentation, location, conditions of use, time.

Retention sample: Sample which is stored for a specified period in case of a need for re-evaluation of data obtained from the main laboratory samples.

Risk: Probability of any defined hazard occurring from exposure to a pesticide under specific conditions. Risk is a function of the likelihood of exposure and the likelihood to harm biological or other systems.

Risk assessment: Process of defining the risk associated with a specified use pattern for a pesticide, usually expressed as a numerical probability or as a margin of safety. Quantifying risk ideally requires,

1. identification of hazard,
2. establishment of dose-response relationships in likely target individuals and populations,
3. exposure assessment (using likely exposure patterns as opposed to worst-case estimates).

Risk management: Decision-making process and procedures used by regulators and others to limit potential risks from use of pesticides. This involves risk assessment, emission control, exposure control and evaluation of the success of the risk mitigation efforts.

Run-off:

1. Movement of a pesticide from a treated field by surface water and eroding sediment.
2. Loss of formulation off foliage during spray application, particularly at high volume.

S

Safener: A substance added to a pesticide formulation to eliminate or reduce phytotoxic effects of the pesticide to certain crops.

Sample: Portion of material selected from a larger quantity of material so that it is representative of the whole. See also aggregate sample, aliquot, composite sample, control sample, increment sample, laboratory sample, primary sample, random sample, retention sample, subsample, test portion and test sample.

Sampling plan: Predetermined procedure for the selection, withdrawal, preservation, transportation, and preparation of the portions to be removed from a population as samples.

SFC: Supercritical fluid chromatography.

SFE: Supercritical fluid extraction.

Soil partition coefficient (K_d): Experimental ratio of a pesticide's concentration in the soil to that in the aqueous (dissolved) phase at equilibrium. It is valid only for the specific concentration and solid/ solution ratio of the test. The K_d is a distribution coefficient reflecting the relative affinity of a pesticide for adsorption by soil solids and its potential for leaching movement through soil.

Soil incorporation: Application of a pesticide to soil by mixing or injection into the soil body.

Soil organic partition coefficient (Koc): Ratio of a pesticide concentration sorbed in the organic matter component of soil or sediment to that in the aqueous phase at equilibrium. The Koc is calculated by dividing the K_d value by the fraction organic carbon present in the soil or sediment.

Soil organic matter: Organic fraction of the soil, including both fresh and aged residues (e.g. humus) of biological origin. Organic carbon refers to that portion of the soil measured as carbon in organic forms, and the organic matter content of soil is assumed to be approximately 1.72-fold that of the organic carbon content.

Sorption: Removal of pesticide from solution by soil or sediment via mechanisms of adsorption and absorption.

SPE: Solid phase extraction.

Specimens: Samples collected from a system for examination, analysis, or storage.

Spiked sample (fortified sample): Control sample with a known amount of pesticide added. Used to test the accuracy (especially the efficiency of recovery) of an analytical method.

Standard solution, primary: Standard prepared by dissolving a weighed amount of an analytical standard pesticide in a known volume of solvent.

Standard solution, secondary: Standard prepared by dilution of an aliquot of a primary standard solution with a known volume of solvent, or by subsequent serial dilutions or a standard solution measured by reference to a primary standard solution.

Sticker: Formulant which increases the adhesiveness of a formulation applied to a surface.

Storage stability test: For a pesticide formulation, a test which measures the chemical and physical stability of the product stored under defined, often worst case, conditions. For pesticide residues, a test which measures stability of residues in stored analytical samples, usually held under frozen conditions at a specified temperature.

Subsample: 1. Portion of the sample obtained by selection or division,
2. Individual unit of the lot taken as part of the sample,
3. Final unit of multistage sampling.

Surfactant: A formulant for reducing interfacial tension of two boundary surfaces, thereby increasing the emulsifying, spreading, dispersability or wetting properties of liquids or solids.

Surface or effective residue: The portion of the insecticide left from the initial deposit is called surface or effective residues. As soon as the deposited material (not adhering tenaciously to the plant or other treated surface) sloughs off, the remaining portion of initial deposit becomes surface or effective residue.

Surveillance: Systematic sampling and residue analysis of commodities, and collation and interpretation of data, in order to ensure compliance with established MRLs. Surveillance may be directed at domestic, imported or exported commodities.

Suspension concentrate (SC): Formulation in which the active ingredient is in the form of a stable dispersion of fine particles in water or organic liquid.

Synergist: Substance, which, while formally inactive or weakly active, can significantly enhance the activity of the active ingredient in a formulation.

Systemic: A systemic pesticide is capable of being translocated to sites other than where it was absorbed in sufficient quantities to be biologically effective.

T

Target, biological: Any organism, organ, tissue, cell or cell constituent that is subject to the action of a pesticide or its residue.

Technical material: Commercial grade .of the pesticide as it comes from the manufacturing plant comprising the active ingredient and associated impurities. It may also contain small quantities of additives necessary for stability.

Terminal residue: Breakdown products of the pesticides, which are stable and create as many problems as the original compound, are called as terminal residue established for a specified, limited period to enable additional biochemical, toxicological or other data to be obtained as may be required for estimating an acceptable daily intake.

Note: A TADI estimated by the Joint FAO/WHO Meeting on Pesticide Residues normally involves the application of a safety factor larger than that used in estimating an ADI.

Test portion (analytical portion): Sub sample of proper size for a chemical analysis or other test, removed from the test sample.

Test sample (analytical sample): Homogenous sample, prepared from the laboratory sample by mixing, grinding, blending, fine-chopping etc. from which test portions are removed for analysis with minimal sampling error.

Test substance: The pesticide as a chemical substance or mixture which is under investigation in a GLP Study.

Test system: Each system (animal, plant, microbial, other cellular, subcellular, chemical, or physical or a combination thereof) used in a study.

Theoretical plates: After the GLC analysis of the known pesticide sample, peak areas are calculated for the comparison of known concentration of chemical with the peak areas. To calculate the peak areas or theoretical plates, the following formula is used:

$$\text{Peak area} = \frac{1}{2}\ b \times h,$$

where b = base of the peak, h = height of the peak

Theoretical maximum daily intake (TMDI): A prediction of the maximum daily intake of a pesticide residue, based on the assumption of levels of residues in food at maximum residue limits and average daily consumption of food per person. The TMDI is expressed in milligrams of residue per person.

Threshold: Concentration of a pesticide in an organism or environmental compartment below which an adverse effect is not expected.

TID: Thermionic detector

TLC: Thin layer chromatography

Tolerable daily intake: Term preferred by the European Commission for acceptable daily intake of environmental contaminants. ADI is reserved for pesticides and food additives where extensive toxicological test data is available.

Tolerance levels: Tolerance levels may be defined as the maximum levels of pesticide residues that are legally permitted, on any agricultural produce or food item when offered for sale or consumption. These tolerance or maximum legal levels for pesticides were prescribed by US EPA, based on a very conservative set of assumptions including that each pesticide is applied at the maximum rate and maximum number of applications with the maximum permissible interval allowed between applications.

Tolerance limit: It is the maximum concentration (in ppm) of pesticide residue that is permitted in or on food at a specified stage of harvesting, transport, marketing or preparation of food up to the final point of consumption. The term now has been replaced by MRL.

Total diet study: Pesticide residue monitoring to establish the pattern of residue intake by a person consuming a defined diet. Primary sampling is as for a market basket survey but the samples are further processed as for domestic consumption i.e. further trimming and cooking as appropriate to local practice.

Total terminal residue: Summation of levels of all the compounds comprising residues of a pesticide in a food.

Transformation product: Chemical species resulting from environmental or metabolic processes on a pesticide.

Translocation: Movement of a substance within the test system or organism.

Transpiration: Vaporisation of water from a leaf into the air.

Treated solution: Test solution that has been subjected to reaction or separation procedures prior to measurement of some property.

Trophic level: Functionally similar organisms such as algae and plants as primary producers are grouped into trophic levels based on similarities in the patterns of food production and consumption.

U

Ultra low volume (ULV) spray: Signifies that the total volume rate of spray application is very low (5 litres per hectare or less). ULV pesticide formulations are generally specially developed for the purpose and are applied undiluted.

Uncertainty factor: Factor in toxicological assessment for extrapolation of data from experimental animals to man (assuming that man may be more sensitive) or from selected individuals to the general population. For example an uncertainty factor is generally applied to the no-observed effect leve1 to derive an acceptable daily intake.

V

Validation: In pesticide analysis, the process for establishing that an analytical method or equipment will provide reliable and reproducible results.

Volatilisation: Evaporation of a pesticide into the atmosphere from a solid or liquid form.

W

Water dispersible granule (WG): Formulation containing granules which readily disperse in water to form a suspension.

Water dispersible powder (WP): Pesticide in a dry form with surfactant, often mixed with or coated on, a fine solid carrier, for dispersion in water to form a suspension.

Water soluble powder: Powder formulation to be applied as a true solution of active ingredient after mixing with water but which may contain insoluble inert ingredients.

Wetting agent: Surfactant for use in spray formulations to assist dispersion of a powder in the diluent or spreading of spray droplets on surfaces. May also incorporate functions of a sticker.

Withholding period: Minimum permissible time between the last application of a pesticide to a crop (including pasture) and harvesting for human consumption or grazing with livestock. The minimum permissible time between the final application of a pesticide to an animal and the collection of eggs or milk or slaughter, for human consumption.

X

Xenobiotic substance: Natural or synthetic compound which is present in an organism but is not a natural component of that organism. Common usage is for man-made environmental contaminants in organisms.

Xylem: Part of the plant's vascular system adapted to the transport of water and solutes from the roots to aerial parts.

Z

Zero tolerance: Limit for a pesticide residue in food or feed which is assumed to be zero and therefore any detectable residue is deemed illegal. Zero tolerances are used by some regulatory systems, e.g. USA, where no maximum residue limits have been established for particular pesticide/ crop combinations.

Reference

Holland, P.T., 1996. Glossary of terms relating to pesticides. *Pure & Appl. Chem.* 68(5): 1167-1193.

Index

A

B

C

D

E

F

G

Zeitfracht Medien GmbH
Ferdinand-Jühlke-Straße 7
99095 Erfurt, Deutschland
produktsicherheit@kolibri360.de